REFORESTING ThE SOUL

OTHER BOOKS BY ANDREW D. MAYES

spiritualityadviser.com

Celebrating the Christian Centuries (1999)
Spirituality of Struggle: Pathways to Growth (2002)
Spirituality in Ministerial Formation (2009)
Holy Land? Challenging Questions from the Biblical Landscape (2011)
Beyond the Edge: Spiritual Transitions for Adventurous Souls (2013)
Another Christ: Re-envisioning Ministry (2014)
Learning the Language of the Soul (2016)
Journey to the Centre of the Soul (2017)
Sensing the Divine (2019)
Gateways to the Divine: Transformative Pathways of Prayer from the Holy City of Jerusalem (2020)
Diving for Pearls: Exploring the Depths of Prayer with Isaac the Syrian (2021)
Voices from the Mountains: Forgotten Wisdom for a Hurting World from the Biblical Peaks (2021)
Climate of the Soul: Ecological Spirituality for Anxious Times (2022)

Reforesting the Soul

Meditating with Trees

ANDREW D. MAYES

RESOURCE *Publications* • Eugene, Oregon

REFORESTING THE SOUL
Meditating with Trees

Copyright © 2022 Andrew D. Mayes. All rights reserved. Except for brief quotations in critical publications or reviews, no part of this book may be reproduced in any manner without prior written permission from the publisher. Write: Permissions, Wipf and Stock Publishers, 199 W. 8th Ave., Suite 3, Eugene, OR 97401.

Resource Publications
An Imprint of Wipf and Stock Publishers
199 W. 8th Ave., Suite 3
Eugene, OR 97401

www.wipfandstock.com

PAPERBACK ISBN: 978-1-6667-5969-3
HARDCOVER ISBN: 978-1-6667-5970-9
EBOOK ISBN: 978-1-6667-5971-6

November 9, 2022 9:19 AM

Unless otherwise acknowledged, Scripture quotations are from New Revised Standard Version Bible, Copyright © 1989, 1995 National Council of the Churches of Christ in the United States of America. Used by permission. All rights reserved worldwide.

Authorized (King James) Version (AKJV) reproduced by permission of Cambridge University Press, the Crown's patentee in the UK.

Complete Jewish Bible (CJB) Copyright © 1998 by David H. Stern. All rights reserved.

God's Word Translation (GW) Copyright © 1995, 2003, 2013, 2014, 2019, 2020 by God's Word to the Nations Mission Society. All rights reserved.

Jerusalem Bible (JB) Copyright © 1966 Darton, Longman & Todd.

The Living Bible Copyright © 1971 by Tyndale House Foundation.

The Message (MSG) Copyright © 1993, 2002, 2018 by Eugene H. Peterson.

The Holy Bible, New Century Version (NCV) Copyright © 2005 by Thomas Nelson, Inc.

New King James Version® Copyright © 1982 by Thomas Nelson. Used by permission. All rights reserved.

New Living Version (NLV) Copyright © 1969, 2003 Barbour Publishing, Inc.

The Passion Translation® Copyright © 2017, 2018, 2020 by Passion & Fire Ministries, Inc. Used by permission. All rights reserved. ThePassionTranslation.com.

The New Testament in Modern English by J.B Phillips copyright © 1960, 1972 J. B. Phillips. Administered by The Archbishops' Council of the Church of England. Used by Permission.

Revised Standard Version (RSV) Copyright © 1946, 1952, and 1971 National Council of the Churches of Christ in the United States of America. Used by permission.

The Voice Bible (VOICE) Copyright © 2012 Thomas Nelson, Inc.

Contents

Photos | vi

Photos—credits | vii

Introduction | ix

1 Deforestation and Replanting, Physical and Spiritual | 1
2 Mystic Metaphors and Sacramental Eyes | 18
3 Jesus and Trees: Secrets of the Kingdom | 28
4 Planting a Fig Tree: Discovering Expectant Silence | 36
5 Planting an Oak: Celebrating the Journey | 47
6 Planting a Willow: Letting Go, Moving On | 63
7 Planting a Vine: Abiding, Fruiting | 76
8 Planting an Apple Tree: Re-entering Eden | 90
9 Planting a Cedar: Standing Tall and Ready | 105
10 Planting an Olive: Rediscovering the Cross | 115
11 Planting a Yew Tree: Regenerating | 127

Bibliography | 141

Photos

1. Horse Chestnut Tree at Vine Cottage, Leominster
2. Trees in Judean Desert
3. Virgin and Child under an Apple Tree by Lucas Cranach
4. Traherne Window "The Tree of Life" by Tom Denny, Hereford Cathedral
5. Beneath a Fig Tree
6. Verdun Memorial Oak, Leominster
7. Harps hung on Willow
8. Beneath the Vine
9. Beneath the Apple Tree
10. "Cedars of God" reserve, Mount Lebanon
11. Planting an Olive
12. Yew tree at Leominster Priory

Photos—credits

1. author
2. author
3. Wikimedia Commons
4. Permission received from the photographer, James Davies
5. author
6. author
7. author
8. author
9. author
10. Wikimedia Commons (uploaded by DIMSFIKAS to https://commons.wikimedia.org/wiki/Category:Cedrus_libani_var._libani#/media/File:The_Cedars_of_God,_Lebanon_2002.jpeg)
11. author
12. author

Horse Chestnut Tree at Vine Cottage, Leominster

Introduction

THE YEAR IS 1161.

His eyes were drawn to wooded crest of Dinmore Hill, just three miles away, majestically dominating the landscape. In times past a vast ancient oak wood had stretched to the Welsh borders and beyond, but now this was the last remaining remnant of a once great forest. As he surveyed the desolate wasteland fields that stretched from Leominster Priory southwards in the direction of the River Arrow at Ivington, Prior Robert was organizing a reclamation of long-neglected terrain. "Well this is a start!" he exhaled, with a deep breath. It was time to bring some renewal to forgotten lands.

Tall and imposing, he bore a certain gravitas and authority that made novices and professed monks alike quake in their boots. "Brother, hold that rod straight!" he barked as he supervised the project with watchful eye and wry smile. The hapless monk Edmund was nervously marking out the burgage plots on the fields south of Leominster's east street, so that the Benedictine community could better manage their lands for the benefit of local people. The first monastic community had been set up by Celtic missionaries from Lindisfarne, led by Edfrith in 660 and it served the district in Anglo-Saxon years as a minster church. Local king Merewalh gifted the lands south of the minster to Edfrith in the 660s as an endowment to resource the work of the church. When a Benedictine Priory was established in the twelfth century by King Henry I, son of William the Conqueror, there was a fresh start for the long-neglected landscape. "At long last this precious land can be used for good," exclaimed the exhilarated Prior. Over the plot Brother Edmund clumsily wielded a perch, the basic unit of measurement for delineating the burgages, five yards in length. The land being enclosed by fences was a strip 200 foot long and 40 feet wide. It had a frontage onto the main street and was to be rented out for fruit growing and vegetable production at a penny a month. The narrow strip of rich soil would sustain good harvests for the poorer families of Leominster.

***

The year is 1751.

The lad was so excited. He had been longing for this day—the chance to escape the two cramped damp rooms which his family had endured for years. The new house was finally complete and the day had come for the little family to move in. It was a modest but handsome timber-framed construction: wattle and daub had now set within the oak posts, creating two rooms and a spacious loft accessed by outside wooden stairs. The long garden was so inviting. For 600 years it had been cultivated intensively with all manner of herbs, fruits and delicious vegetables that had even supplied the local market just yards away on Corn Square Market. Now, it was beginning to morph into a fruit-filled orchard and woodland glade: the vegetable plots were beginning to give way to trees, as tenants' energy turned to local trade—at this time Etnam Street was beginning to fill with all manner of businesses and workshops. "I really want to mark this day!" Will cried out to his father. "I've had this conker in my pocket for ages. I got in the woods near Hereford. Where can I plant it, Father? We have arrived!"

***

The year is 2019.

I gasped out loud when I first saw the majestic tree. The horse chestnut towers more than one hundred feet high, with a wide crown and vast canopy that can be seen from quite a distance. I later discovered its roots—coming up under the linoleum in the kitchen extension which the Victorians had built—making the tree less than forty feet from the house. In spring its candles are ablaze, while in autumn it's risky to walk under it—so many fresh conkers are falling to the ground!

I had been drawn initially to the cottage by its name "Vine Cottage." Sure enough the vine crept up the wall which the Georgians built to consolidate the burgage plot. But I was soon enchanted by the long garden—it formed an Eden, a paradise filled now with trees of all kinds including yew, rowan, conifer, beech, elderberry, apple, walnut, hawthorn, goat willow, magnolia, purple smoketree. On the boundary thrive laurel and silver birch. We have made our own additions—near the fruit bushes (raspberry, blueberry, blackberry, raspberry, redcurrant) we have planted a little orchard of pear, plum, peach, cherry—and even added, as an act of faith, olive trees, a lemon tree and a Japanese maple.

The burgage is a veritable ecosystem, with amazing biodiversity: squirrels playing mischievously in the branches, noisily snuffling hedgehogs, frogs, honey bees and all manner of birdlife: wood pigeons and turtle doves, chaffinches and bluetits, wrens, robins and blackbirds, swallows dancing in the sky overhead at dusk. But it was the experience of planting the trees that set my imagination afire, leading to this book, first used by retreatants here at Borderlands Retreats. For me it became a parable of the spiritual life. Where there had been bare earth or overgrowing briars now there is fruitfulness and fecundity.

From our back gate, opening onto Sydonia Park, we like Prior Robert enjoy a view of Dinmore Hill. It had been utterly decimated in the First World War when requisition of timber—needed for everything from fuel, to shoring up trenches, lining them with duckboards, pitprops and even ammunition boxes—caused widespread tree felling and the denudation of the once heavily-forested mount. But the woodlands were to see a revival. Once called Queen's Coppice in honor of Elizabeth I they became Queenswood when an arboretum was established in 1953 to mark the coronation of Queen Elizabeth II. Now it is home to a stunning tree collection covering 67 acres with 1,200 rare and exotic trees from all over the world and 103 acres of semi-natural ancient woodland, which is designated as a Site of Special Scientific Interest (SSSI).

As I reflect on the transformation of Dinmore Hill and our own modest replanting here at Vine Cottage I am reminded of the imagery used by the prophets . . .

DEFORESTATION AND RENEWAL

The prophets describe in vivid, arresting terms the state of a ruinous countryside. This imagery serves two purposes: it represents the threat of an invader, the fear of a coming judgement, and, at the same time, expresses the desolation of the people's soul.

Isaiah depicts a deforestation and denuding of the land:

> The earth shall be utterly laid waste and utterly despoiled;
> The earth dries up and withers,
>> the world languishes and withers;
>> the heavens languish together with the earth.

> The earth lies polluted
> > under its inhabitants. (Isa 24:3-5)

> For the fortified city is solitary,
> > a habitation deserted and forsaken, like the wilderness;
> the calves graze there,
> > there they lie down, and strip its branches.
> When its boughs are dry, they are broken;
> > women come and make a fire of them.
> For this is a people without understanding. (Isa 27:10,11)

At the call of the prophet Isaiah and his big-hearted response "Here am I; send me!" he is given a fearsome message:

> Astonished, I said, "And Master, how long is this to go on?" He said, "And even if some should survive, say a tenth, the devastation will start up again. The country will look like pine and oak forest with every tree cut down— Every tree a stump, a huge field of stumps. (Isa 6:13)

However there is a glimmer of hope, for the message continues immediately:

> But there's a holy seed in those stumps. (Isa 6:13, *Message* translation)

Later, Isaiah's successor will declare the glorious promise of God:

> I will make the wilderness a pool of water,
> > and the dry land springs of water.
> I will put in the wilderness the cedar,
> > the acacia, the myrtle, and the olive;
> I will set in the desert the cypress,
> > the plane and the pine together,
> so that all may see and know,
> > all may consider and understand,
> that the hand of the Lord has done this,
> > the Holy One of Israel has created it. (Isa 41:18-20).

The prophet celebrates a divine replanting—an inspiring image that encourages us to glimpse the possibilities in our own spiritual life for reforesting the arid soul.

The prophet Joel—writing in the eighth century BC—traces a similar journey from deforestation to renewal. He begins with a picture of a desolate landscape of the soul:

INTRODUCTION

> [The enemy] has laid waste my vines,
>   and splintered my fig trees;
> it has stripped off their bark and thrown it down;
>   their branches have turned white . . .
> The fields are devastated,
>   the ground mourns;
> for the grain is destroyed,
>   the wine dries up,
>   the oil fails.
> Be dismayed, you farmers,
>   wail, you vinedressers,
> over the wheat and the barley;
>   for the crops of the field are ruined.
> The vine withers,
>   the fig tree droops . . .
> To you, O Lord, I cry.
> For fire has devoured
>   the pastures of the wilderness,
> and flames have burned
>   all the trees of the field. (1: 7, 10-12, 19)

However, penitence opens the people to a new beginning:

> Yet even now, says the Lord,
>   return to me with all your heart,
> with fasting, with weeping, and with mourning;
>   rend your hearts and not your clothing.
> Return to the Lord, your God,
>   for he is gracious and merciful,
> slow to anger, and abounding in steadfast love,
>   and relents from punishing.
> Who knows whether he will not turn and relent,
>   and leave a blessing behind him? (2:12-14)

Joel announces the promise of renewal and speaks directly to the soil:

> Do not fear, O soil;
>   be glad and rejoice,
>   for the Lord has done great things!
> Do not fear, you animals of the field,
>   for the pastures of the wilderness are green;
> the tree bears its fruit,
>   the fig tree and vine give their full yield. (2:21-22)

And he declares words that Peter will pick up at Pentecost:

> Then afterward
> I will pour out my Spirit on all flesh;
> your sons and your daughters shall prophesy,
> your old men shall dream dreams,
> and your young men shall see visions.
> Even on the menservants and maidservants,
> in those days, I will pour out my Spirit. (2:28,29)

Streams of restorative grace will once again flow through the devastated land.

Amos tells us that he is a worker with trees before he is a prophet:

> Then Amos answered Amaziah, "I am no prophet, nor a prophet's son; but I am a herdsman, and a dresser of sycamore trees, and the Lord took me from following the flock, and the Lord said to me, 'Go, prophesy to my people Israel.' (Amos 7: 14-15)

He finds himself struggling for social justice against the exploitation and oppression of the vulnerable—the powerful were selling off needy people for goods, taking advantage of the helpless, oppressing the poor. Drunk on their own economic success and intent on strengthening their financial position, the people had lost the concept of caring for one another. Amos rebuked them because he saw in that lifestyle evidence that Israel had forgotten God. As one who worked closely with trees, Amos turned naturally to the imagery of physical desolation and the hope of renewal. Blessings will return if the people heed the fundamental call: "Let justice roll down like waters, and righteousness like an ever-flowing stream" (5:24).

> The time is surely coming, says the Lord,
> when the one who plows shall overtake the one who reaps,
> and the treader of grapes the one who sows the seed;
> the mountains shall drip sweet wine,
> and all the hills shall flow with it.
> I will restore the fortunes of my people Israel,
> and they shall rebuild the ruined cities and inhabit them;
> they shall plant vineyards and drink their wine,
> and they shall make gardens and eat their fruit.
> I will plant them upon their land,
> and they shall never again be plucked up
> out of the land that I have given them,
> says the Lord your God. (Amos 9: 13-15)

INTRODUCTION

THE AIM OF THIS BOOK

This book explores pathways to renewal through the powerful metaphor of reforesting the desert places. The soul can sometimes be an arid, thirsty desiccated place, becoming as exhausted and denuded as land that has been ravaged and a stripped of its trees. God's promise is to reforest the wilderness and renew our fruitfulness. This book is a guided retreat, resonating with our contemporary and urgent ecological concerns—and the shift from egocentrism to ecocentrism. It explores the rich symbolism of different trees both in the Bible and in the Christian tradition, including hymnody and poetry. As land that is reforested holds the promise of rejuvenation and new beginnings, so this book heartens us with pointers towards spiritual rejuvenation.

The need to address the task of reforestation in the natural world is urgent and vital. Mark Kinver, Environment reporter for BBC News writes:

> The world is losing the battle against deforestation. A historic global agreement aimed at halting deforestation has failed, according to a report. An assessment of the New York Declaration on Forests (NYDF) says it has failed to deliver on key pledges. Launched at the 2014 UN climate summit, it aimed to half deforestation by 2020, and halt it by 2030. Yet deforestation continues at an alarming rate and threatens to prevent the world from preventing dangerous climate change, experts have said.
>
> The critique, compiled by the NYDF Assessment Partners (a coalition of 25 organizations), painted a bleak picture of how the world's forests continue to be felled. "Since the NYDF was launched, deforestation has not only continued—it has actually accelerated," observed Charlotte Streck, co-founder and director of Climate Focus, which coordinated the publication of the report.
>
> The report says the amount of annual carbon emissions resulting from deforestation around the globe is equivalent to the greenhouse gases produced by the European Union. On average, an area of tree cover the size of the United Kingdom was lost every year between 2014 and 2018.[1]

But there are signs of hope.

The 2020 World Economic Forum, held in Davos, announced the creation of the ambitious *Trillion Tree Campaign*, aiming to plant 1 trillion trees across the globe. The Nature Conservancy's *Plant a Billion Trees*

---

1. Mark Kinver, "World Losing Battle against Deforestation."

initiative is a major forest restoration effort with a goal of planting a billion trees across the planet. On a more modest—but achievable—scale BBC's Countryfile launched in 2020 a project to plant 750,000 trees. The two-year initiative, called *Plant Britain*, encourages viewers to grow the trees at various sites across the UK in an effort to tackle climate change. The Queen's Green Canopy was a tree planting initiative created to mark Elizabeth II's Platinum Jubilee in 2022, everyone across the UK invited to plant trees to create a network of individual trees, avenues, copses and whole woodlands in honor of the Queen's service and the legacy she built. The Woodland Trust expressed its hope: "This will create a green legacy of its own, with every tree planted bringing benefits for people, wildlife and climate, now and for the future."[2] And at the COP26 climate summit, held in Glasgow in November 2021, more than 100 world leaders promised to end and reverse deforestation by 2030, a pledge costed at almost $20bn.

THE CHALLENGE AND THE INVITATION

All this becomes a parable for the spiritual life. Like land that is despoiled through deforestation many experience the soul as overworked and underfed, exhausted and eroded. In addition to usual stresses and strains, the pandemic has been devastating both for people's mental health and for the state of the soul. Many feel uprooted spiritually, their soul like a landscape scarred and laid waste.

The time has come for a re-planting in the soul, that we might thrive again and release life-giving oxygen of love to a hurting, breathless world. In the face of our seeming powerlessness, we can be decisive. We *can* take action. The key issue we will face is: *what does it mean to plant this tree in my soul?* To plant a tree in our soul is to make a commitment, to forge an act of intentionality, to reclaim our freedom, take control and take responsibility for our spiritual lives. It is to tell ourselves: with God's empowering grace, we *can* do something creative, positive, even life-changing. Under the pandemic we have lived in a state of subservient passive resignation for too long. Take your spade into your hand! The time has come to start digging!

What difference will it make? We will see that each tree has its singular charism and particular challenge. Pertinent scriptures are examined. Spiritual writers from the Christian tradition through the centuries will help us

2. Woodland Trust, "Queen's Green Canopy."

INTRODUCTION

unpack these. Penetrating questions throughout and suggested prayer exercises make this a transformational journey—towards an Eden regained!

SUMMARY OF THIS BOOK

We begin by looking at the dangers and opportunities represented in deforestation and replanting, physical and spiritual. The perilous state of the world's forests today gives us clues for the reading of our own soul, while Biblical writers clarify for us dangers and possibilities in the spiritual life. Gerard Manley Hopkins helps us give voice to our fears and hopes.

In chapter two, we learn to interpret the world playfully, enquiringly, differently—using mystic metaphors and sacramental eyes. Prime examples encourage us: while Ephrem invites us to see God's symbols everywhere, Hildegard of Bingen calls us to a greening of the soul.

Next we take a look at how Jesus communicates his Good News through the metaphor of the tree, and how trees can reveal "the secrets of the Kingdom of God."

In chapter four Nathaniel under his fig tree beckons us to open up spaces in our lives where we can come and rest a while and hear God's call to us in stillness and silence. Newly-discovered writings by seventh century Isaac the Syrian help us reflect on this theme.

Next we see how the oaks invite us to celebrate God's providence and guidance in our life-pilgrimage thus far. John Bunyan shares his experience of walking in the way of the patriarchs.

In chapter six we explore how the willow allows us to "hang up our harps" and let go of attachments and burdens, and identify possibilities for rejoicing. Having released our grip on the past, De Caussade guides us towards the sacrament of the present moment.

Then, the vine calls us to stillness and to abide in God's grace, while grapes crushed summon us towards adventurous ministry. We explore this paradox of settling and serving with the help of Francis, Julian, Teresa and Benedict.

In chapter eight the apple tree not only summons us back to Eden and forwards to heaven but invites us to delight fully in Christ's grace here and now. Poets from the fourth, seventeenth and eighteenth centuries—Ephrem, Traherne and Hutchins—are our guides.

Next we allow the cedar to invite us to celebrate our lofty dignity, and to consider our availability to others. Contrasting traditions hearten us: Baptist Charles Spurgeon and Trappist Thomas Merton.

In the succeeding chapter the olive challenges us to live the life-giving paradox of the Cross. We are inspired in our response by the Fortunatus, the epic poem Dream of the Rood, and by Bonaventure.

Finally, the yew tree invites us to take stock of our spiritual life and holds out to us possibilities of unending renewal, encouraging us to develop a strategy for ongoing rejuvenation. John Henry Newman invites us to continual openness to God's call and beckonings in our life.

As we begin this retreat-journey we pray for an attentiveness to the trees around us. We seek an openness and readiness to allow the trees to question us and pose their challenges to us. We pray for the grace to replant our soul, opening us to renewal and new beginnings.

TREES IN JUDEAN DESERT

1 Deforestation and Replanting, Physical and Spiritual

Your country lies desolate . . .
　I will put in the wilderness the cedar,
　　the acacia, the myrtle, and the olive (Isa 1:7; 41:18)

PHYSICALITY POINTS TO SPIRITUALITY. The physical landscape points to what Gerard Manley Hopkins called the "inscape" of the soul. Surveying the outer terrain helps us read the contours of our soul—it gives us clues and pointers: taking in a perspective opens up for us insights and challenges for the spiritual life.

DEFORESTATION ACROSS THE WORLD TODAY

Forests cover a third of the planet's land surface, providing food, medicine and fuel for more than a billion people. The precious resource of forests are regarded by some as large, undeveloped swathes of land that can be converted for purposes such as agriculture and grazing. In North America, half the forests in the eastern part of the continent were cut down for timber and farming between the 1600s and late 1800s.

Today, most deforestation happens in the tropics. Areas once inaccessible are now within reach as new roads are ploughed through the dense forests. In 2017 alone the tropics lost about 61,000 square miles of forest—an area the size of Bangladesh.

The World Bank estimates that four million square miles of forest have been lost since the beginning of the 20th century. In this century alone, forests shrank by half a million square miles— an area bigger than the size of South Africa. In 2018, The *Guardian* reported that every second, a chunk of forest equivalent to the size of a football pitch is lost.

1 DEFORESTATION AND REPLANTING, PHYSICAL AND SPIRITUAL

Deforestation occurs when a forested area is cut and cleared to make way for agriculture, grazing or timber. Human-lit fires clear land for agricultural use. First, valuable timber is harvested, then the remaining vegetation is burned to make way for crops like soy or for cattle grazing. In August 2019 alone, for example, more than eighty thousand fires burned in the Amazon, an increase of almost 80% from 2018.

Many forests are cleared to make way for plantations of palm oil, the most commonly produced vegetable oil, found in half of all supermarket products, from food to shampoo.

Forests are home to 80% of terrestrial biodiversity, containing a wide array of trees, plants, animals and microbes. Deforestation in tropical regions affects the way water vapor is produced over the canopy, reducing rainfall and exacerbating drought. Breathing out oxygen, trees absorb carbon dioxide, alleviating the sun-blocking effect of greenhouse gas emissions. Trees play a crucial role in the capture of excess carbon dioxide. Tropical trees alone are estimated to provide about a quarter of the climate mitigation needed to offset climate change, while deforestation accounts for nearly 20% of greenhouse gas emissions. The lungs of the world, trees are needed more than ever before in the face of unrelenting climate change.

PHYSICAL DEFORESTATION IN THE SCRIPTURES

The Hebrew scriptures give us an insight into the deforestation of the land of Israel and environs and its consequences. Four themes emerge.

1. Grasping Ambition

The first Israelite settlers entering the land of Canaan under Joshua's leadership looked in envy at those already living in the land. It seems that the fertile valleys were already occupied by the Canaanites, and the tribes were left with the only option of settling in the rugged hill country, densely forested in places. They complained about this and Joshua was forced to concede that they must cut down forests in order to achieve their goals of settlement:

> The tribe of Joseph spoke to Joshua, saying, "Why have you given me but one lot and one portion as an inheritance, since we are a numerous people, whom all along the Lord has blessed?" And Joshua said to them, "If you are a numerous people, go up to the

forest, and clear ground there for yourselves in the land of the Perizzites and the Rephaim, since the hill country of Ephraim is too narrow for you." The tribe of Joseph said, "The hill country is not enough for us; yet all the Canaanites who live in the plain have chariots of iron, both those in Beth-shean and its villages and those in the Valley of Jezreel." Then Joshua said to the house of Joseph, to Ephraim and Manasseh, "You are indeed a numerous people, and have great power; you shall not have one lot only, but the hill country shall be yours, for though it is a forest, you shall clear it and possess it to its farthest borders; for you shall drive out the Canaanites, though they have chariots of iron, and though they are strong." (Josh 17:14–18)

Environmental scientist Daniel Hillel observes:

> Clearing the forest in the hill district required much work and risked land degradation. The dense growth of trees and shrubs that once covered the slopes had protected the soil naturally. Removing that cover and pulverizing the ground by cultivation as well as allowing livestock to trample it, made the soil vulnerable to accelerated erosion, which could be controlled partially by the painstaking construction of walled terraces.[1]

Deuteronomy (20:19) gives this advice:

> When thou shalt besiege a city a long time, in making war against it to take it, thou shalt not destroy the trees thereof by forcing an axe against them: for thou mayest eat of them, and thou shalt not cut them down, for the tree of the field is man's life. (AV)

As the NLV puts it:

> . . . you must not cut down the trees with your axes. You may eat the fruit, but do not cut down the trees. Are the trees your enemies, that you should attack them?

2. Exploitation by Developers

A second cause of deforestation was the ambition of builders demanding excessive amounts of timber for their grandiose constructions. It was said of Solomon: "God gave Solomon very great wisdom, discernment, and breadth of understanding . . . He would speak of trees, from the cedar that

---

1. Hillel, *Natural History*, 300n14.

1 DEFORESTATION AND REPLANTING, PHYSICAL AND SPIRITUAL

is in the Lebanon to the hyssop that grows in the wall" (1 Kgs 4:29, 33). As we shall consider in chapter ten, Solomon's building of the temple was a priority. But he did not stop there. In fact, he caused the massive felling of cedar trees in Lebanon, floated down the Mediterranean coast on rafts:

> Solomon was building *his own house* for thirteen years, and he finished his entire house. He built *the House of the Forest of the Lebanon* one hundred cubits long, fifty cubits wide, and thirty cubits high, built on four rows of cedar pillars, with cedar beams on the pillars. It was roofed with cedar on the forty-five rafters, fifteen in each row, which were on the pillars. There were window frames in the three rows, facing each other in the three rows. All the doorways and doorposts had four-sided frames, opposite, facing each other in the three rows. He made *the Hall of Pillars* fifty cubits long and thirty cubits wide. There was a porch in front with pillars, and a canopy in front of them. He made *the Hall of the Throne* where he was to pronounce judgement, *the Hall of Justice*, covered with cedar from floor to floor.
>
> His own house where he would reside, in the other court behind the hall, was of the same construction. Solomon also made *a house like this hall for Pharaoh's daughter*, whom he had taken in marriage ... There were costly stones above, cut to measure, and cedar wood. The great court had three courses of dressed stone to one layer of cedar beams all round; so had the inner court of the house of the LORD, and the vestibule of the house. (I Kgs 7: 1–8, 11, 12)

Solomon was not content with the one building of the temple—he wanted to develop a complex at the southern end of the Ophel Ridge in Jerusalem, the so-called city of David. In particular he built another substantial structure, rivalling the temple in splendor and timber-usage, which by its very name acknowledged its debt to the forests of Lebanon:

> He made 300 shields of beaten gold, using three hundred shekels of gold on each shield, and the king put them in the House of the Forest of Lebanon. (2 Chr 9:16)

The "House of the Forest of Lebanon", a monumental military building in Jerusalem is regarded as an armory within the palace because of the references to shields. But Lebanon paid a heavy price for fulfilling Solomon's personal ambitions. Hillel observes:

> In ancient times, the entire mountain range of Lebanon was covered with a dense growth of cedars, cypresses, and junipers.

5

The Phoenicians began to cut the forest trees systematically, for constructing boats and houses, and—being astute traders—they exported the timber. Over the ensuing centuries, the forests were gradually cleared and the denuded mountainsides exposed to erosion.[2]

The temple was destroyed by Babylonians in 586BC and began to be rebuilt by returning exiles in 516. The prophet Haggai responds to the command: " Go up to the hills and bring wood and build the house" (1:8). In about 444BC Nehemiah organized work on the walls around it. He had become one of the most important men in Persia (modern Iran): a cupbearer to Artaxerxes, king of Persia, and later, the king made him a governor of Persian Judea. So he dared to make a substantial request for sourcing timber from Persian forests:

> Then I said to the king [Artaxerxes], "If it pleases the king, let letters be given me to the governors of the province Beyond the River, that they may grant me passage until I arrive in Judah; and a letter to Asaph, the keeper of the king's forest, directing him to give me timber to make beams for the gates of the temple fortress, and for the wall of the city, and for the house that I shall occupy." And the king granted me what I asked, for the gracious hand of my God was upon me. (Neh 2:8)

We note that Nehemiah sought this timber not only for city walls, but for his own rather nice house!

3. Misuse of Resources

Several of the prophets condemn the extraction of wood—not for reasons of ecological degradation, but because of the perverted use of the timber. They declare in scathing terms that the forests were not given them by God for such use:

> All who make idols are nothing, and the things they delight in do not profit; their witnesses neither see nor know. And so they will be put to shame. Who would fashion a god or cast an image that can do no good? Look, all its devotees shall be put to shame; the artisans too are merely human. Let them all assemble, let them stand up; they shall be terrified, they shall all be put to shame . . .

---

2. Hillel, *Natural History*, 172.

> The carpenter stretches a line, marks it out with a stylus, fashions it with planes, and marks it with a compass; he makes it in human form, with human beauty, to be set up in a shrine. He cuts down cedars or chooses a holm tree or an oak and lets it grow strong among the trees of the forest. He plants a cedar and the rain nourishes it. Then it can be used as fuel. Part of it he takes and warms himself; he kindles a fire and bakes bread. Then he makes a god and worships it, makes it a carved image and bows down before it. Half of it he burns in the fire; over this half he roasts meat, eats it and is satisfied. He also warms himself and says, "Ah, I am warm, I can feel the fire!" The rest of it he makes into a god, his idol, bows down to it and worships it; he prays to it and says, "Save me, for you are my god!"
>
> They do not know, nor do they comprehend; for their eyes are shut, so that they cannot see, and their minds as well, so that they cannot understand. No one considers, nor is there knowledge or discernment to say, "Half of it I burned in the fire; I also baked bread on its coals, I roasted meat and have eaten. Now shall I make the rest of it an abomination? Shall I fall down before a block of wood?" He feeds on ashes; a deluded mind has led him astray, and he cannot save himself or say, "Is not this thing in my right hand a fraud?" (Isa 44:9- 20)

Jeremiah echoes Isaiah about the misuse of God-given forest resources:

> For the customs of the peoples are false:
> a tree from the forest is cut down,
>   and worked with an ax by the hands of an artisan;
> people deck it with silver and gold;
>   they fasten it with hammer and nails
>   so that it cannot move.
> Their idols are like scarecrows in a cucumber field,
>   and they cannot speak;
> they have to be carried,
>   for they cannot walk.
> Do not be afraid of them,
>   for they cannot do evil,
>   nor is it in them to do good. (Jer 10:3–5)

4. Images of Judgement

As we have already seen in the writings of Joel, Amos and Isaiah, the destruction of beautiful forests becomes an icon of the degradation that the people bring on their own soul. Placed on the lips of Moses are the devastating words:

> The next generation, your children who rise up after you, as well as the foreigner who comes from a distant country, will see the devastation of that land ... all its soil burned out by sulphur and salt, nothing planted, nothing sprouting, unable to support any vegetation, like the destruction of Sodom and Gomorrah. (Deut 29:22-23)

In a later generation the prophet Zechariah will share his heartache:

> Open your doors, O Lebanon,
>   so that fire may devour your cedars!
> Wail, O cypress, for the cedar has fallen,
>   for the glorious trees are ruined!
> Wail, oaks of Bashan,
>   for the thick forest has been felled!
> Listen, the wail of the shepherds,
>   for their glory is despoiled!
> Listen, the roar of the lions,
>   for the thickets of the Jordan are destroyed! (Zech 11:1-3)

Isaiah clearly used the destruction of the forest as an image or metaphor of coming judgement on Assyria:

> And He will destroy the glory of his forest and of his fruitful garden,
> both soul and body,
> And it will be as when a sick man wastes away.
> And the rest of the trees of his forest will be so small in number
> That a child could write them down. (10:18-19)

Those who boast proudly of the deforestation they have achieved reveal not only their arrogance and spiritual pride but also their contempt against the Creator:

> Whom have you mocked and reviled?
>   Against whom have you raised your voice
> and haughtily lifted your eyes?
>   Against the Holy One of Israel!
> By your servants you have mocked the Lord,

and you have said, 'With my many chariots
I have gone up the heights of the mountains,
    to the far recesses of Lebanon;
I felled its tallest cedars,
    its choicest cypresses;
I came to its remotest height,
    its densest forest.
I dug wells
    and drank waters,
I dried up with the sole of my foot
    all the streams of Egypt.' (Isa 37:23–25)

Jeremiah in like fashion uses the image as both metaphor and physical reality:

> The word that the LORD spoke to the prophet Jeremiah about the coming of King Nebuchadrezzar of Babylon to attack the land of Egypt . . .
> her enemies march in force,
> and come against her with axes,
>   like those who fell trees.
> They shall cut down her forest,
> says the LORD,
>   though it is impenetrable,
> because they are more numerous
>   than locusts;
>   they are without number.
> Daughter Egypt shall be put to shame;
>   she shall be handed over to a people from the north. (Jer 46:13, 22–24)

While we recall these four themes of deforestation in the Bible, let us not lose sight of Job's words:

> For there is hope for a tree,
>   if it is cut down, that it will sprout again,
>   and that its shoots will not cease.
> Though its root grows old in the earth,
>   and its stump dies in the ground,
> yet at the scent of water it will bud
>   and put forth branches like a young plant. (Job 14: 7–9)

LEARNING FROM CONTEMPORARY DEFORESTATION

While the biblical examples we have encountered give us clues for reading our own soul, so too we find striking parallels between present-day physical degradation of the land and the spiritual dangers we face in the practice of ministry and discipleship. Five things stand out:

1. Exhaustion

Today in many places across the globe we see evidence that once rich soil has become overworked. Persistent degradation of dryland ecosystems is exacerbated by overgrazing: in places where forests once flourished animals now eat away at grasses and topsoil becomes damaged by their hooves until very little goodness is left. This resonates with a sense of stress and burnout experienced by clergy or lay workers who too often harassed by the demands put upon them. They may be taken for granted not appreciated or affirmed. Some face debilitating fatigue when working a seventy hour week. Wildfires and flare ups in the natural world mirror episodes of stress which can contribute to spiritual breakdown.[3] Energy and spiritual reserves become depleted.

2. Starvation

Today we see with our own eyes how denuded and starved land can become where deforestation has led to unsustainable farming depleting the nutrients in the soil. So too in our own lives we sometimes realize that we are not feeding the soul or drawing on the needful spiritual nutrients that nourish the inner life.

3. Desiccation

In the wake of deforestation climate change has led to increasing droughts and heat waves—these are everywhere projected to become more intense. Temperatures are expected to continue rising, with a corresponding reduction of soil moisture. Timber becomes tinder as wildfires ignite among bone-dry trees. In our own spiritual lives we too face times of aridity and

---

3. See, for example, Society of Martha and Mary, *Affirmation and Accountability*.

dryness of soul. We thirst for something more. We long for our spiritual drought to be inundated, saturated and drenched by the waters of the Spirit.

4. Erosion

Desiccated landscapes across the planet experience both wind erosion and flash flooding, aggravating the damage, carrying away topsoil and leaving behind an infertile mix of dust and sand. Clear-cutting of land, when the tree and plant cover that binds the soil is removed, and the stripping away of trees for fuel and timber or to clear land for cultivation—it is the combination of these factors that transforms degraded land into desert. This gives us a powerful image of the vulnerability of soul we sometimes face—we may feel being exposed, worn down, and even our sense of calling may become eroded.

5. Exploitation

On many continents we have witnessed how the removal of trees has been motivated by a consumerist drive for different products. We have seen how the original purpose and vocation of forests has been overturned by a dramatic change of use—for example where palm trees are planted after the removal of native trees, or animal grazing occurs where natural forest once thrived. Those in ministry also experience changing goalposts and unrealistic expectations put upon them. Clergy complain that they have become managers instead of pastors—management of plant and finances, meeting quotas, and team-management, together with imposed bureaucracy, has displaced the joy of pastoral encounters and led some to notice both the use and abuse of ministers.

We echo the lament of Gerard Manley Hopkins (1844–89) in his poem *Binsey Poplars*, felled 1879:

> My aspens dear, whose airy cages quelled,
> > Quelled or quenched in leaves the leaping sun,
> > All felled, felled, are all felled;
> > > Of a fresh and following folded rank
> > > > Not speared, not one
> > > > That dandled a sandalled
> > > > > Shadow that swam or sank
> > > On meadow and river and wind-wandering weed-

> winding bank.
> O if we but knew what we do
>   When we delve or hew—
>   Hack and rack the growing green!
> After-comers cannot guess the beauty been,
>   Ten or twelve, only ten or twelve
>   Strokes of havoc unselve
>   That sweet especial scene,
>   Rural scene, a rural scene,
>   Sweet especial rural scene.[4]

But Hopkins has opportunities to delight in new beginnings. In *The May Magnificat* he asks:

> Question: What is Spring?
>   Growth in every thing . . .
> When drop-of-blood-and-foam-dapple
> Bloom lights the orchard-apple
>   And thicket and thorp are merry
>   With silver-surfed cherry
> And azuring-over greyball makes
> Wood banks and brakes wash wet like lakes . . .
> This ecstasy all through mothering earth[5]

Winter passes, spring is come. In his *Journal* Hopkins records both sorrow and joy:

> 8 April 1873
>
> The ashtree growing in the corner of the garden was felled. It was lopped first: I heard the sound and looking out and seeing it maimed there came at that moment a great pang and I wished to die and not to see the inscapes of the world destroyed any more . . .
>
> 8 August 1874
>
> In the evening I went by myself up the hills towards Bishopsteignton. I looked over a hedge down to a row of seven slender rich elms at a bottom between two steep fields: the run of the trees and their rich and handsome leafage charmed and held me.[6]

---

4. Hopkins, *Poems*, 39, 40.
5. Hopkins, *Poems*, 38, 39.
6. Hopkins, *Poems*, 128, 132.

1 DEFORESTATION AND REPLANTING, PHYSICAL AND SPIRITUAL

REPLANTING THE SOUL

What is needed for re-foresting, for a spring-time in the soul?

Preparation

> Sow for yourselves righteousness; reap steadfast love; break up your fallow ground; for it is time to seek the LORD, that he may come and rain righteousness upon you. (Hos 10:12)

If the scriptures challenge us on the causes of deforestation physical and spiritual, they also hearten us and encourage us to plant seeds and nurture fresh sprigs. Reforestation, of course, requires careful preparation of the land: the soil might need to be scarified and the debris removed—an image suggestive of the spiritual practice of confession. As in the case of planting of seeds in Jesus' beautifully-observed Parable of the Sower, we need to see how the soil of our soul might become more receptive. How would you describe the soil of your spiritual life? The parable speaks of "rocky ground, where the seeds did not have much soil, and they sprang up quickly, since they had no depth of soil" (Matt 13:5). It notices how in land with briars "the thorns grew up and choked the seeds" (13:7). This is land facing competition between species that struggle to get the mastery—and the better sunlight. For Jesus they represent "the cares of the world and the lure of wealth" which "choke the word, and it yields nothing" (13:22). Good soil is described as "one who hears the word and understands it, who indeed bears fruit" (13:23). At the outset of this process, we pray for self-knowledge and clarity about the state of our own soul, and our relationship with God. It is helpful to survey our own spiritual landscape and notice what this imagery suggests to us.

Synergy

The scriptures suggest that such a planting is a divine-human partnership:" I planted the seed. Apollos watered it, but it was God Who kept it growing. For we work together with God" (1 Cor 3:6,9, NLV). We must take responsibility for the state of our soul and take initiatives. But only with divine grace, for God plants within the soul: we become "oaks of righteousness, the planting of the Lord" (Isa 61). And so we rejoice with the prophet:

> Shout for joy, O heavens, for the Lord has done it!
> Shout joyfully, you lower parts of the earth;
> Break forth into a shout of joy, you mountains,
> O forest, and every tree in it;
> For the Lord has redeemed Jacob
> And in Israel He shows forth His glory. (Isa 44:23)

FROM DEATH TO LIFE

> My beloved speaks and says to me:
> "Arise, my love, my fair one,
>   and come away;
> for now the winter is past,
>   the rain is over and gone.
> The flowers appear on the earth;
>   the time of singing has come" (Song 2:10–12)

The passage from spiritual de-forestation to replanting and renewal evokes the paschal theme of dying—to live. After times when our spiritual experience can be depicted in terms of barrenness, dryness, dying we can, with God's grace, experience periods when our prayer can be described as spiritual springtime: budding flourishing, blossoming—growth and fruitfulness. Seeds that looked unpromising become planted, germinating, fruiting. This evokes the Easter hymn "Now the green blade riseth from the buried grain": seed falling into the ground and dying to enable a great harvest (John 12:24):

> When our hearts are wintry, grieving, or in pain,
> Thy touch can call us back to life again,
> Fields of our hearts that dead and bare have been:
> Love is come again, like wheat that springeth green.[7]

D. H. Lawrence (1885–1930) in his poem *Shadows* hopes for snatches of renewal

> odd, wintry flowers upon the withered stem, yet new, strange flowers
> such as my life has not brought forth before, new blossoms of me[8]

---

7. J.M.C. Crum (1872–1958).
8. Lawrence, *Complete Poems*.

1 DEFORESTATION AND REPLANTING, PHYSICAL AND SPIRITUAL

"For everything there is a season, and a time for every matter under heaven" (Eccl 3:1). There will be times and seasons in the spiritual life: the winter and desolation of the soul give way to re-energizing springtime. John of Damascus (d.754) puts it in his Easter song:

> All the winter of our sins,
> Long and dark is flying
> From his light . . .
> Now the queen of seasons, bright
> with the day of splendor[9]

Richard Rolheiser reminds us that we can experience "deaths" in one of two ways. We can treat something as a "terminal death", a death that ends life and ends possibilities; a bleak finality. Or we can experience a dying as a "paschal death" which, while marking the end of one pattern of living, opens us to new and potentially richer ways of life. [10]

> His shoots shall spread out; his beauty shall be like the olive tree, and his fragrance like that of Lebanon. They shall again live beneath my shadow, they shall flourish as a garden; they shall blossom like the vine, their fragrance shall be like the wine of Lebanon. (Hos 14:5–7)

As we begin the journey of this retreat, we pray with the psalmist for the grace of planting fresh trees in the soul that will bear a great harvest:

> Happy are those
> [whose] delight is in the law of the LORD,
>   and on his law they meditate day and night.
> They are like trees
>   planted by streams of water,
> which yield their fruit in its season,
>   and their leaves do not wither.
> In all that they do, they prosper (Ps 1)

PRAYER EXERCISE

In the silence, reflect on:

---

9. *Come, ye faithful, raise the strain*, tr. J. M. Neale, (1818–66).
10. Rolheiser, *Seeking Spirituality*, ch. 7.

1. Which of the four causes of physical deforestation speak to you the most?
2. Which of the five features of deforestation resonate most strongly with your present experience?
3. As you reflect on the promises of God about replanting, what is your greatest longing and need at the beginning of this retreat?
4. Taking a walk in your environs, look attentively for any signs in the trees or natural world of decay and renewal. Especially search for paschal evidences of fresh growth amidst dyings off.

VIRGIN AND CHILD UNDER AN APPLE TREE BY LUCAS CRANACH

2   Mystic Metaphors and Sacramental Eyes

The heavens are telling the glory of God (Ps 19:1)

COMMUNICATING THE MYSTERIES OF THE SOUL

How can we describe to others what is happening to us on our spiritual journey? How can we depict, for the benefit of ourselves and for others, the spiritual road that we are taking: experiences of prayer, transitions that we travel through, impediments that we face? Barry and Connelly put it: "most people are inarticulate when they try . . . to describe their deeper feelings and attitudes. They can be even less articulate when they try to describe their relationship with God . . . For to begin to talk about this aspect of their lives requires the equivalent of a new language, the ability to articulate inner experience."[1]

Campbell tells us: "Here we sense the function of metaphor that allows us to make a journey we could not otherwise make."[2] As we seek to bring to expression aspects of our inner, spiritual life, we discover that we need metaphors, frameworks, reference points. Jurgen Moltmann affirms how vital it is to use images to describe spiritual experience: "In the mystical metaphors, the distance between a transcendent subject and its immanent work is ended . . . the divine and human are joined in an organic cohesion."[3]

---

1. Barry and Connolly, *Spiritual Direction*, 67.
2. Campbell, *Thou art That*, 9.
3. Moltmann, *Spirit of Life*, 285.

Metaphors Enable us to be Spiritual Explorers

Paul Avis points out that metaphors drawn from the natural world are used by poets as a hermeneutical key to help map the landscapes of the mind: "Metaphor is generated in the drive to understand experience ... Metaphor is not just naming one thing in terms of another, but seeing, experiencing and intellectualizing one thing in the light of the other."[4] Brian Wren puts it: "Metaphors can ... extend language, generate new insights, and move us at a deep level by their appeal to the senses and imagination."[5] Metaphors have the power to shift us from left-brain analytical thinking to creative right-hemisphere imagining—and imaging.

In his study *The Edge of Words: God and the Habits of Language* Rowan Williams summons us to be unhesitating in our use of metaphors: "So as we take more risks and propose more innovations in our linguistic practice, we move from the more-or-less illustrative use of a vivid and unusual simile through to increasingly explosive usages that ultimately ... invite us to rethink our metaphysical principles, our sense of how intelligible identities are constructed in and for our speaking. Extreme or apparently excessive speech is not an aberration in our speaking."[6]

Metaphors stoke and trigger the imagination. Janet Martin Soskice affirms their cognitive role, aiding and deepening our understanding of things: "what is said by the metaphor can be expressed adequately in no other way ... the combination of parts in a metaphor can produce new and unique agents of meaning."[7] Metaphors evoke and stimulate rather than define or confine. Metaphors help to unify and integrate experience, because they link the spiritual to the physical, and the soul to the body, enabling the metaphysical to become physical. There is a potential in metaphor properly described as *heuristic:* the word means "stimulating further investigation, encouraging discovery through experimenting, exploration something by first-hand experience." We need to rediscover the sacramentality of words: like bread and wine they can bear God's presence and reveal the Divine—so words should be approached with reverence and appreciation. Soskice affirms: "The sacred literature ... both records the

---

4. Avis, *Creative Imagination*, 97.
5. Wren, *What Language*, 92.
6. Williams, *Edge of Words*, 130.
7. Soskice, *Metaphor*, 31. See also Lakoff and Johnson, *Metaphors We Live By*.

experiences of the past and provides the descriptive language by which any new experience may be interpreted."[8]

LOOKING AT THE WORLD WITH SACRAMENTAL EYES

A sacramental way of viewing reality is a dominant theme in the fourth gospel. Jesus sees wine, vines, water, bread, sunlight and candlelight, even shepherding, as speaking of himself. Jesus looks at a seed and sees its potential if it perishes: "Unless a grain of wheat falls into the earth and dies, it remains just a single grain; but if it dies, it bears much fruit" (12:24). He glimpses his very destiny in a kernel of wheat.

The other gospels combine to give us the clear impression that this was an outlook on the world that was truly characteristic of Jesus himself. The secrets of the Kingdom reveal themselves through parables of seed, mountain, field and sea (Matt 13; Mark 11:23). Jesus says: "Consider the lilies, how they grow" (Luke 12:27). "Consider": the Greek word means "turn your attention to this, notice what is happening, take note." It is a summons to a contemplative way of life, a deeply reflective way of seeing the world. John 4 gives some examples.

Seeing Differently

While in Samaria, Jesus encounters a fixation with physical things and leads his hearers into a radically different way of looking at the world.

Twice over, the woman at the well is stuck on a literal and physical hearing of Jesus' words: "Sir, you have no bucket, and the well is deep. Where do you get that living water? . . . Sir, give me this water, so that I may never be thirsty or have to keep coming here to draw water" (4:11,15). But Jesus wants to lead her from the physicality of the water to its sacramentality, and how it powerfully symbolizes the gift of God. The physical water and the well speak to Jesus of humanity's deep thirst for things of the Spirit and God's gracious provision.

The disciples too are utterly bewitched by a concern for physical things. They had gone into the town to buy food (4:8). Upon their return, they urge him: "Rabbi, eat something" (4:31). But he said to them, "I have

---

8. Soskice, *Metaphor*, 160. See also Mayes, *Language of the Soul*.

food to eat that you do not know about." So the disciples said to one another, "Surely no one has brought him something to eat?'" (4:32–33).

Jesus sees food as highly symbolic and sacramental. In chapter 6, John will put on the lips of Jesus: "The bread that I will give for the life of the world is my flesh" (6:51). There, he will be misunderstood and accused of advocating cannibalism. Another response will be to ask for a continual supply of free, fresh bread (6:34). His hearers stay on the level of the physical and cannot glimpse sacramentality. In Samaria, he explains: "My food is to do the will of him who sent me" (4:34). He is not talking about a picnic brought to him. He is talking of the deep nourishment and sustenance that come from moving within the Father's will.

Jesus wants to open the disciples to a new vision and a fresh way of seeing reality. He calls them to become wide awake to the possibilities God is opening up: "Look around you, and see how the fields are ripe for harvesting" (4:35). But he is not talking about Samaritan agriculture. The fields around them speak to Jesus of the growth of the Kingdom and the spiritual harvest that has become imminent: "I tell you, lift up your eyes" (4:35, RSV).

This sacramental way of looking at the world stands in utter contrast with the way the woman and the disciples see things—they can't see past a bucket of water or a plate of food! "Take a look around you," Jesus says to the disciples. Learn to see things differently.

Glimpsing the Divine

The poem or hymn of creation found in the opening lines of Genesis celebrates the goodness of God's world: "The earth brought forth vegetation: plants yielding seed of every kind, and trees of every kind bearing fruit with the seed in it. And God saw that it was good" (Gen1:12). Through the centuries poets and artists communicate spiritual realities through vivid imagery from flora and fauna, discovering natural, horticultural, agricultural and temporal metaphors for the spiritual life. Gerard Manley Hopkins puts it:

> The world is charged with the grandeur of God.
> It will flame out, like shining from shook foil[9]

William Blake in his poem *Auguries of Innocence* invites us

9. Phillips (ed.), *Gerard Manley Hopkins*.

> To see a world in a grain of sand
> And a heaven in a wild flower,
> Hold infinity in the palm of your hand,
> And eternity in an hour.[10]

Troubadour St Francis of Assisi, the first poet to compose in the Italian vernacular, invites us to recognize and celebrate the radical and essential interconnectedness of all things, displaying a remarkable kinship and sense of unity with creation in his *Canticle of Creation*. He hailed the sun as brother and the moon as sister; he greeted Sister Water and Brother Wind, and in his ministry he approached the fearsome wolf of Gubbio as "brother." At the dawn of capitalism and a creeping consumerist approach to things—Francis was the son of a wealthy cloth-merchant and worked in his shop—he discovered a deep connectedness to all things which was honoring and non-exploitative. Franciscan prayer nurtures such an appreciative and respectful approach to the world of nature.[11] The natural world points to the new creation: the seen evokes the unseen, earth points to heaven. We learn to delight in images that organically spring from the world of nature.

The Greening of Soul

As a prime example both of enjoying metaphors and of sacramental seeing, Hildegard of Bingen (1098–1179), poet, mystic and musician, celebrates our "greening" or *viriditas*. Today we talk about the "greening of the planet" but nine hundred years ago Hildegard celebrated the presence of the Holy Spirit in the created order through the idea of greening. Indeed, Hildegard herself invented the word *viriditas*. She talks about the "exquisite greening of trees and grasses" and goes on: "the earthly expression of the celestial sunlight; greenness is the condition in which earthly beings experience a fulfilment which is both physical and divine; greenness is the blithe overcoming of the dualism between earthly and heavenly."[12] For Hildegard, the wetness or moisture of the planet, revealed in verdant growth, bespeaks the Holy Spirit who "poured out this green freshness of life into the hearts of men and women so that they may bear good fruit."[13] In another place she

---

10. William Blake, from notebooks now known as *The Pickering Manuscript*.
11. See, for example, Stoutzenberger and Bohrer, *Praying with Francis of Assisi*.
12. Peter Dronke, quoted in Bowie and Davies (eds.), *Hildegard of Bingen*, 32.
13. Peter Dronke, quoted in Bowie and Davies (eds.), *Hildegard of Bingen*, 32.

writes: "Greening love hastens to the aid of all. With the passion of heavenly yearning, people who breathe this dew produce rich fruit."[14]

She invites us to see the world differently, overcoming the dichotomy of heaven and earth by glimpsing the heavenly action in the freshness of the planet, which mirrors the human soul.

For Hildegard, the greatest sin is the sin of drying up. While moisture and wetness bespeak the work of the Holy Spirit, spiritual aridity is the telltale sign of our thirst for such grace and divine transformation:

> If we surrender the green vitality of virtues and give ourselves over to the drought of our indolence so that we lack the sap of life and the greening power of good deeds, then the powers of our very soul will begin to fade and dry up. [15]

> When a forest does not green vigorously,
> Then it is no longer a forest.
> When a tree does not blossom,
> It cannot bear fruit.
> Likewise a person cannot be fruitful
> Without the greening power of faith,
> And an understanding of scripture. [16]

> Envy drives out all greening power! . . .

> Lies are like juiceless foam,
> Hard and black,
> Lacking the verdancy of Justice,
> Dry,
> Totally without tender goodness . . .
> Now in the people
> That were meant to green,
> There is no more life of any kind.
> There is only shriveled barrenness. [17]

But fecundity and radiance come to those open to be re-greened:

> Good People,
> Most royal greening verdancy,

---

14. Fox, *Illuminations of Hildegard of Bingen*, 43.
15. Fuhrkotter, *Hildegardis Scivias*, 473.
16. Uhlein, *Hildegard of Bingen*, 62.
17. Uhlein, *Hildegard of Bingen*, 75–77.

Rooted in the sun,
You shine with radiant light . . .
God hugs you.
You are encircled
By the arms
Of the mystery of God.[18]

Hildegard casts fresh light on an enigmatic saying of Jesus:

> "For if men use the green wood like this, what will happen when it is dry?" (Luke 23:31). Jesus himself was the green wood because he caused all the greening power of the virtues. Yet he was rejected by unbelievers. The Antichrist, however, is the dry wood because he destroys all the living freshness of justice and cases things that should be green to wither away.[19]

Gloriously, she hails a saint as

> Life-giving greenness of God's hand!
> Through you, God has planted an orchard . . .
> You rise resplendent[20]

Even in England, in their own way, poets echo this theme. Poet-priest George Herbert asks in *The Flower*:

> Who would have thought my shrivelled heart
>   Could have recovered greenness?

And in his poem *Thou art indeed just, Lord* Gerald Manley Hopkins prays: "O thou Lord of life, send my roots rain."

Looking Afresh at the World

We might learn from the astonishing sacramental worldview of Ephrem (306–373).

> In every place, if you look, His symbol is there. (Hymns on Nativity 21:6)
> Lord, Your symbols are everywhere
> Blessed is the Hidden One shining out. (On Faith. 4:9)[21]

---

18. Uhlein, *Hildegard of Bingen*, 90.
19. Fox, *Divine Works*, 243.
20. "Song in Honor of St Disibodo" in Fox, *Divine Works*, 390.
21. Brock, *Luminous Eye*, 39.

2 MYSTIC METAPHORS AND SACRAMENTAL EYES

The Syriac tradition encourages us to read the two books of nature and Scripture. Ephrem affirms:

> The keys of doctrine
> which unlock all of Scripture's books,
> have opened up before my eyes
> the book of creation,
> the treasure house of the Ark,
> the crown of the Law.
> This is a book which, above its companions,
> has in its narrative
> made the Creator perceptible
> and transmitted His actions;
> it has envisioned all His craftsmanship,
> made manifest His works of art. (Hymns on Paradise 6:1,25)
>
> In his book Moses described the creation of the material world,
> so that both Nature and Scripture might bear witness to the Creator:
> Nature, through man's use of it, Scripture, through his reading of it.
> These are the witnesses which reach everywhere,
> they are to be found at all times, present at every hour.
> (Hymns on Paradise 5:2)[22]

In this retreat we learn with Hildegard and Ephrem to see how the whole created order brims with the Divine and teaches us about God's ways. This is not a utilitarian approach to the natural world, looking around for helpful illustrations or analogies for the spiritual life. Rather, it is a question of training ourselves to recognize the revelatory character of creation and how God teaches us through it.

PRAYER EXERCISE

In contemporary times, with smart phones and Ipads, we are often one step removed from the natural world, disconnected, literally losing touch with it. So explore with inquisitive eye your immediate environs. What do you notice? Celebrate the details of your surroundings, environment, setting, context.

The Franciscan Bonaventure (13C) encourages us to both appreciate the smallest, tiniest features on earth and also the magnitude and vastness of creation in sky and cloud. He says:

22. Brock, *Hymns on Paradise*, 108–9, 118, 102.

> The beauty of things in the variety and light, shape and color in simple mixed or organic bodies—such as heavenly bodies and minerals like stones and metals, and plants and animals—clearly proclaims the divine power, wisdom and goodness.

Recapture a sense of curiosity. Learn to be intrigued. Give thanks for the incarnate God, the God of the cosmos and the God of the detail.

Ask yourself: where am I glimpsing God, or feeling his presence? What things speak to me of the Divine? Maybe read your environs symbolically and look for elements that somehow represent the sort of God you believe in. What metaphors of the soul emerge for you?

Perhaps you can sketch a picture or draw a symbol or write a poem to express your discoveries?

As you open up your awareness and consciousness of the Divine, conclude with thanksgiving.

TRAHERNE WINDOW "THE TREE OF LIFE" BY TOM DENNY, HEREFORD CATHEDRAL

3   Jesus and Trees

Secrets of the Kingdom

"Lift up your eyes!" (John 4:35)

JESUS ENCOURAGES US TO look to the trees for vital clues about our spiritual life. As a carpenter, he worked reverentially with timber. In Galilee he was surrounded by lushly wooded hills, while in Jerusalem Jesus cherished a particular grove of olive trees at the foot of the Mount of Olives, the Garden of Gethsemane. The gospels tell us that this was a place Jesus often made a retreat. Luke writes: "He came out and went, as was his custom, to the Mount of Olives; and the disciples followed him" (Luke 22:39) while John notes: "he went out with his disciples across the Kidron valley to a place where there was a garden . . . Jesus often met there with his disciples" (John 18:1,2). The Garden of Gethsemane is a woodland grove at the foot of the mountain opposite the walled city of Jerusalem. It is close to the city but also affords a distance, both physical and reflective, from the hustle and bustle, from the controversies and conflicts. From this viewpoint Jesus could look up at the temple on the Ophel Ridge across the Kidron Valley, its marble and gold surfaces opposite gleaming and glistening in the sunlight. The olive trees here gave Jesus a space in which to think and pray. Their shade offered Jesus a place of solitude and solace away from the crowds, as well as an undisturbed spot where he could talk privately with his disciples.

Jesus instructs his disciples: "Lift up your eyes" (John 4:35). In Galilee he does just that: "consider the flowers of the field . . . consider the birds of the air" (Matt 6:26–34). Luke tells us that in Jericho "When Jesus came to the place, he looked up"—seeing Zacchaeus in the sycamore tree he says: "The Son of Man came to seek out and to save the lost" (19:10). Jesus continues his journey to Jerusalem to be greeted with waving palms and his very destiny culminates on a tree: "We are witnesses to all that he did both

3 JESUS AND TREES

in Judea and in Jerusalem. They put him to death by hanging him on a tree" (Acts 10:39). Throughout his ministry, Jesus delights in the metaphor of the tree, giving vital teaching with reference to fig, mulberry, mustard and vine trees. In fact, he welcomes the opportunity to let trees speak to us about seven key themes:

1. Trees reveal the Kingdom of God growing from small beginnings

> He also said, "With what can we compare the kingdom of God, or what parable will we use for it? It is like a mustard seed, which, when sown upon the ground, is the smallest of all the seeds on earth; yet when it is sown it grows up and becomes the greatest of all shrubs, and puts forth large branches, so that the birds of the air can make nests in its shade. (Matt 13:31–32)

Jesus is saying two things in this parable. Firstly, the reign of God may begin in seemingly insignificant, unimpressive, unpromising ways, like a tiny seed. But it will become visible and expansive. Do not underestimate the beauty and the potential of the small. Do not rule anything out or give in to fatalism and resignation. Nothing is inevitable, little is predictable. Do not despise the little things. The germ of an idea can flourish into a philosophy. The imperative is to sow seeds in trust, hope and utter openness to God's future. He has great things in store for us that we can't begin to glimpse now. As Ephesians (3:20,21) puts it:

> Now to him who by the power at work within us is able to accomplish abundantly far more than all we can ask or imagine, to him be glory in the church and in Christ Jesus to all generations, for ever and ever. Amen. (NRSV)

> God can do anything, you know—far more than you could ever imagine or guess or request in your wildest dreams! He does it not by pushing us around but by working within us, his Spirit deeply and gently within us.
>
> Glory to God in the church!
> Glory to God in the Messiah, in Jesus!
> Glory down all the generations!
> Glory through all millennia! Oh, yes! (Eph 3:20,21 *Message*)

The reforesting of the soul begins with small decisions and a dose of faith.

Secondly, within the Kingdom of God there is a space for all. Its branches are both far-reaching and all-inclusive. They welcome those looking for a home, as birds seek safe spaces where they can build nests. The Kingdom of God can embrace and enfold all.

2. Trees Reveal to us the State of the Soul

> You will know them by their fruits. Are grapes gathered from thorns, or figs from thistles? In the same way, every good tree bears good fruit, but the bad tree bears bad fruit. A good tree cannot bear bad fruit, nor can a bad tree bear good fruit. Every tree that does not bear good fruit is cut down and thrown into the fire. Thus you will know them by their fruits. (Matt 7:16–20)
>
> Either make the tree good, and its fruit good; or make the tree bad, and its fruit bad; for the tree is known by its fruit. (Matt 12:33)

The repetition of this saying "by their fruits ye will know them" indicates that this is a key observation by Jesus with an important message. The outer evidence of our lives gives indicators and clues as to the state of our soul. Paul spells this out plainly:

> Now the works of the flesh are obvious: fornication, impurity, licentiousness, idolatry, sorcery, enmities, strife, jealousy, anger, quarrels, dissensions, factions, envy, drunkenness, carousing, and things like these . . . By contrast, the fruit of the Spirit is love, joy, peace, patience, kindness, generosity, faithfulness, gentleness, and self-control . . . If we live by the Spirit, let us also be guided by the Spirit. (Gal 5:19–23,25)

Modern translations of Matthew 7:16–20 bring out the clear message of Jesus:

> If you grow a healthy tree, you'll pick healthy fruit. If you grow a diseased tree, you'll pick worm-eaten fruit. The fruit tells you about the tree. (*Message*)
>
> You must determine if a tree is good or rotten. You can recognize good trees by their delicious fruit. But if you find rotten fruit, you can be certain that the tree is rotten. The fruit defines the tree. (*Passion* translation).

What does your life taste like? We pray for self-knowledge as we reflect on our attitudes, actions, reactions, priorities and motives—what do they tell us about the state of our relationship with God?

3. Trees Speak of Patient Nurturing

> Then he told this parable: "A man had a fig tree planted in his vineyard; and he came looking for fruit on it and found none. So he said to the gardener, 'See here! For three years I have come looking for fruit on this fig tree, and still I find none. Cut it down! Why should it be wasting the soil?' He replied, 'Sir, let it alone for one more year, until I dig around it and put manure on it. If it bears fruit next year, well and good; but if not, you can cut it down.'" (Luke 13:6–9)

Refuse fatalism, and do not be too quick to pass judgement on your own planting. "Let it alone for one more year, until I dig around it and put manure on it." Any planting or replanting needs both not only determination in the grower but also consistent care and feeding. We cannot expect tender and vulnerable shoots to flourish in our life without ensuing their protection and nurture. The two characters in this parable take opposite stands. The owner wants a quick return but may know nothing about horticulture and the fruiting of trees. The gardener is characterized by wisdom and patience, and does not expect quick results. Planting and replanting in the soul need dedicated care. We must be gentle with ourselves. The important thing is to make the most of the present moment, to seize the opportunity God gives us and not rush to judgment. "One more year"—God is patient with us and often gives us a second chance, another growing season, even if we do not deserve it or expect it.

4. Trees Invite us to Expectant Faith

Jesus draws our attention to trees in order to encourage us to have big faith. First there is the saying about a mustard seed and mulberry tree:

> The apostles said to the Lord, "Increase our faith!" The Lord replied, "If you had faith the size of a mustard seed, you could say to this mulberry tree, 'Be uprooted and planted in the sea,' and it would obey you." (Luke 17:5,6)

This message is developed and expanded in reference to a fig tree in Holy Week:

> On the following day, when they came from Bethany, he was hungry. Seeing in the distance a fig tree in leaf, he went to see whether perhaps he would find anything on it. When he came to it, he found nothing but leaves, for it was not the season for figs. He said to it, "May no one ever eat fruit from you again." And his disciples heard it . . .
>
> In the morning as they passed by, they saw the fig tree withered away to its roots. Then Peter remembered and said to him, "Rabbi, look! The fig tree that you cursed has withered." Jesus answered them, "Have faith in God. Truly I tell you, if you say to this mountain, 'Be taken up and thrown into the sea,' and if you do not doubt in your heart, but believe that what you say will come to pass, it will be done for you. So I tell you, whatever you ask for in prayer, and it will be yours. (Mark 11:12–14, 20–24)

In the first saying Jesus once again speaks about faith beginning small. In the second passage Jesus moves the focus towards the practice of prayer. He strongly emphasizes the setting aside of doubt. The phrase "believe that you have received it" (11:24) appears in some manuscripts as "believe that you *are receiving* it." Jesus is commending a way of praying that is expectant and receptive. It is not a passive resignation before God ("I will leave the results to you"). It is an active praying, on tiptoes, ready to welcome and accept the divine response. This is taught to us by the rather strange way of cursing a barren fig tree. Nevertheless, it is so clear that when Jesus looks at a tree, he recognizes and discerns its message. The trees speak to him. They actually reveal life in the Kingdom.

5. Trees Speak of Jesus' Passion

In addition to this Holy Week episode, trees repeatedly feature in the passion story. Palm trees (John 12:13) assist in Jesus' welcome into the city as suffering servant:

> A very large crowd spread their cloaks on the road, and others cut branches from the trees and spread them on the road. The crowds that went ahead of him and that followed were shouting, "Hosanna to the Son of David!" (Matt 21:8–9)

3   JESUS AND TREES

On the way of the cross Jesus utters an enigmatic saying to the weeping women of Jerusalem:

> For if they do this *when the wood is green*, what will happen when it is dry? (Luke 23:28–31).

We have noted Hildegard's observation on this about Jesus' greening power. The green tree seems to represent Jesus, while the dry tree those pitched against him. It might be understood: "If they do these things in *me*, fruitful, always green, undying, what will they do to *you*, fruitless, and unresponsive?" Perhaps we can best leave it as speaking to us of the sometimes enigmatic and puzzling message of the trees!

As we will consider later, the New Testament often talks of a tree (or of its wood) when referring to the crucifixion:

> The God of our ancestors raised up Jesus, whom you had killed by hanging him on a tree. (Acts 5:30)

Branches or beams now welcome not the birds of the air but the Son of God. They enable him to stretch out his arms to embrace and enfold the whole world.

6. Trees Alert us to the Coming of Kingdom of God

> Then they will see "the Son of Man coming in clouds" with great power and glory . . . From the fig tree learn its lesson: as soon as its branch becomes tender and puts forth its leaves, you know that summer is near. So also, when you see these things taking place, you know that he is near, at the very gates. Truly I tell you, this generation will not pass away until all these things have taken place. Heaven and earth will pass away, but my words will not pass away. "But about that day or hour no one knows, neither the angels in heaven, nor the Son, but only the Father. Beware, keep alert; for you do not know when the time will come. (Mark 13:26, 28–33)

The fig tree once again has its messages. Now its tender shoots, so keenly awaited by the gardener, speak to us of the need for sharpened alertness and vigilance. We should be looking for the signs of the Kingdom as keenly as a gardener is on the look out for the first sign of summer. We should live in a state of heightened expectancy and keep our eyes and hearts wide open. We might miss a sign of the Kingdom under our very nose! What will we spot today of God's presence, and what will we miss?

7. Trees Express our Relationship with God

> I am the true vine, and my Father is the vine-grower . . . Abide in me as I abide in you . . . Those who abide in me and I in them bear much fruit, because apart from me you can do nothing . . . (John 15:1,4)

Chapter seven will help us explore this key metaphor. For the moment let us note how Jesus sees in the vine a clear picture of our relationship to him. He does not communicate this relationship through lectures or abstract ideas, but in a visceral, tactile, even sensual way.[1] If we have eyes to see, let us look! Let us learn to see creation in the way Jesus models for us.

In this introductory overview we recognize that trees can communicate to the open soul what Jesus calls "the secrets of the Kingdom of God " (Luke 8:10). Now it is time for some serious replanting!

PRAYER EXERCISE

Take a walk in nature praying to look at creation through the eyes of Jesus.

Seek an attentiveness and alertness as you pray with your five senses. Pray with your fingertips, touching and caressing the surfaces of trees you encounter—delighting in the texture of their bark. Feel the veins in the leaves. Look closely at the foliage in all its variety on the trees. Enjoy the scents. Is there a fruit you can taste? Listen to the sounds of birdsong and the breeze and breath of the wind through the branches.

Above all, pray for the grace to discover in the trees the secrets of the Kingdom. Does any new parable suggest itself?

End by giving thanks for the diversity of creation and the ability to enjoy God's world.

---

1. See Mayes, *Sensing the Divine*.

BENEATH A FIG TREE

4 Planting a Fig Tree

Discovering Expectant Silence

The fig tree puts forth its figs, and the vines are in blossom; they give forth fragrance.
Arise, my love, my fair one, and come away. (Song 2:13)

The next day Jesus decided to go to Galilee. He found Philip and said to him, "Follow me." Now Philip was from Bethsaida, the city of Andrew and Peter. Philip found Nathanael and said to him, "We have found him about whom Moses in the law and also the prophets wrote, Jesus son of Joseph from Nazareth." Nathanael said to him, "Can anything good come out of Nazareth?" Philip said to him, "Come and see." When Jesus saw Nathanael coming toward him, he said of him, "Here is truly an Israelite in whom there is no deceit!" Nathanael asked him, "Where did you get to know me?" Jesus answered, "I saw you under the fig tree before Philip called you." Nathanael replied, "Rabbi, you are the Son of God! You are the King of Israel!" Jesus answered, "Do you believe because I told you that I saw you under the fig tree? You will see greater things than these." And he said to him, "Very truly, I tell you, you will see heaven opened and the angels of God ascending and descending upon the Son of Man." (John 1:43–51)

1. WHAT IS THE SIGNIFICANCE OF THE FIG TREE?

AT ITS OUTSET, JOHN'S gospel takes us to the fig tree. John is evoking Eden. He has already quoted from Genesis 1 "In the beginning"—and wants to take us to our origins. There were three trees in Eden, the Tree of Life, the Tree of the Knowledge of Good and Evil, and a third: the fig tree.

> The woman took of its fruit and ate; and she also gave some to her husband, who was with her, and he ate. Then the eyes of both were opened, and they knew that they were naked; and they sewed fig leaves together and made loincloths for themselves. They heard the sound of the LORD God walking in the garden at the time of the evening breeze, and the man and his wife hid themselves from the presence of the LORD God among the trees of the garden. But the LORD God called to the man, and said to him, "Where are you?" He said, "I heard the sound of you in the garden, and I was afraid, because I was naked; and I hid myself." He said, "Who told you that you were naked?" (Gen 3: 6–11)

In Genesis, the fig tree is associated with shame, guilt and coverup. Nathaniel, almost like a second Adam, finds himself beneath such a tree, but he is characterized by openness, honesty and truth. Jesus will say to him: "Here is truly an Israelite in whom there is no deceit!" A tree associated with hiding from God now becomes, with the advent of Jesus, a tree that represents a refreshing openness to the Divine.

2. WHAT WAS NATHANIEL DOING UNDER THE FIG TREE?

He is finding stillness

"Nathanael, right before Philip came to you, I saw you sitting under the shade of a fig tree" (1:48, *Passion* translation). Nathaniel is sitting. Movement and activity have ceased. As an outer physical stillness descends on his body, so an inner quietude creeps into his soul. He is at rest. He has stopped running around responding to life's demands. He has come to a place where he can open himself utterly to God.

He is expectant

In the scriptures the fig tree had become a symbol of messianic hope and contentment (see "every man sitting under his fig tree" in Deut 8, Micah 4:4 and Zech 3:10). Sitting under the fig tree signifies messianic peace and becomes symbolic of a yearning for the advent of the redeemer and the messianic age. Shortly he will be hailed by Jesus as a representative Israelite.

He is waiting

Later Jesus himself will see the fig tree as a place of promise. "From the fig tree learn its lesson: as soon as its branch becomes tender and puts forth its leaves, you know that summer is near. So also, when you see all these things, you know that he is near, at the very gates" (Matt 24: 32). The fig tree becomes a place of expectancy and hope.

He reflects

We picture Nathaniel in meditation underneath the tree. The image that presents itself is one who has withdrawn there for thought or prayer. Jewish writings tell of distinguished rabbis who were accustomed to rise early and pursue their studies under the shade of a fig tree. Augustine, in his *Confessions*, relates of himself: "I cast myself down, I know not how, under a certain fig tree, giving full vent to my tears; and the floods of mine eyes gushed out, an acceptable sacrifice to Thee" (viii. 28).

Questions are bubbling up in Nathaniel's soul, which he will later verbalize as he says to Philip, "Can anything Good come out of Nazareth?" Maybe he needs to let go of preconceptions, prejudices and inherited concepts in order to see that God shatters our preconceived ideas of people and places and possibilities. His question reveals the honesty in Nathaniel's heart—he is unafraid to speak out his puzzlement and doubts. The fig tree becomes a place of reflection and honest questioning.

He seeks solitude

Nathaniel has withdrawn from the world of work and activity and made some kind of retreat beneath the fig tree. He was getting away from the crowd and from the duties imposed upon him. As a Pharisee he was normally engaged in noisy debate about the interpretation of the scriptures. Now he craves silence. He seeks to be alone with God. The fig tree becomes a place of prayer and attentiveness to God.

He seeks renewal

Perhaps—we don't know—Nathaniel is utterly wearied and exhausted, in body and soul. From the tree he seeks both shade from the blazing sun and refuge from demands put upon him. He seeks both rest and rejuvenation.

3. WHAT DOES JESUS SAY TO HIM?

"I saw you"

> When Jesus saw Nathanael coming toward him, he said of him, "Here is truly an Israelite in whom there is no deceit!" Nathanael asked him, "Where did you get to know me?" Jesus answered, "I saw you under the fig tree before Philip called you."

These words denote not one isolated glance but Jesus' deep insight into the depths of the soul: past events and deepest longings are known to Jesus. This is not just bare recognition. It is not only that Nathaniel was noticed; it means that he was understood. "Before Philip called you"—before he came out into the open, when he was half hidden, when his soul-life was secret and unconfessed—"I saw you."

Jesus knows him. He understands his heart's longings, desires and hurts. He feels his woundedness and spiritual thirst. John puts it: He "knows what is in man" (2:25, RSV); "No one needed to tell him about human nature, for he knew what was in each person's heart" (NLT). Here is no hiding as in Genesis, no cover up, but rather an exposure of heart to God, a transparency to him who sees all, a nakedness of soul.

Nathaniel is allowed to be himself, and as such, he feels loved, accepted and understood, even though he has uttered few words. The fig tree becomes a place of unfettered encounter with the Divine.

"You will see greater things than these . . . You will see heaven opened"

As the *Voice* translation puts it:

> Nathanael, if all it takes for you to believe is My telling you I saw you under the fig tree, then what you will see later will astound you. I tell you the truth: before our journey is complete, you will

see the heavens standing open while heavenly messengers ascend
and descend, swirling around the Son of Man. (John 1:50,51)

Jesus promises Nathaniel new vision and the gift of revelation. Jesus is offering, on earth, the very vision of heaven!

Throughout the fourth gospel, Jesus is seeking eyes wide open. He begins his ministry with the summons, the invitation: "Come and see" (1:39). He himself is looking acutely: he turns and sees Andrew and Simon following him (1:38) and looking intently at Simon (1:42) he gives him a new name and a new destiny. The invitation "Come and see", first issued by Jesus, is taken up and echoed by Philip (1:46) and by the woman of Samaria (4:29). "Come and look!" From the beginning to the end of the gospel, we are summoned to look, to open our eyes wide. The imperative "behold!":"Take a long look!" is found from the Baptist to Pilate.

John the Baptist cries out: "Look! The Lamb of God . . . !" (1:36). On Palm Sunday the evangelist quotes Zechariah: "Look, your king is coming" (12:15). Even Pilate will say: "Behold the man . . . behold your king!" (19:5,14, AV). We are directed to notice, peer beneath the surface, watch, contemplate, gaze. The challenge, then, of this Gospel is to open wide our eyes, minds and hearts. It is all about looking. "And the Word became flesh and lived among us, and we have seen his glory, the glory as of a father's only son, full of grace and truth" (1:14). Jesus calls us to be curious, inquisitive, enquiring—not only to look, but also to see.

Jesus is inviting Nathaniel to an adventure and an odyssey of the soul, a pilgrimage of faith. We know nothing of what happened to Nathaniel but he shows up at end of the gospel meeting the risen Christ by the Sea of Galilee, where he is named as being among the disciples (John 21:2). The Pharisee became a disciple. The fig tree was a launching pad for an untold journey of conversion and transformation. Who can tell what might happen to us? We can only imagine what transpired for Nathaniel, who like his fellow Pharisee Nicodemus appears at the start and end of the gospel. One thing is certain, the time will come for him to leave his resting place under the fig tree and set out on a journey of a lifetime!

FINDING STILLNESS

Many spiritual writers urge us towards the kind of stillness that prepared Nathaniel for receptivity to the Divine. Isaac the Syrian from the seventh century, whose writings[1] have only just been discovered, has some encouragement to give us. He prays: "immerse my soul in the deepest depths of quiet in God" (1/LXVI:314). Isaac leads us towards a place of stillness where stirrings and turbulence are left behind. Here prayer learns to be receptive to the Divine: "On the level and in the life of the spirit . . . [human nature] remains in a certain and inexplicable silence, for the working of the Holy Spirit stirs in it, it being raised above the realm of the soul's understanding" (2/XXXII:4). Isaac encourages us to sink deeper in our prayers—the deeper we go, the more we advance towards stillness and leave the eddies behind. We are summoned to a depth of stillness beyond surface stirrings and anxieties, resting in God in this inexplicable stillness, becoming a "sharer in the mystery of God" (2/XX:15).

Here Isaac distinguishes between what he calls pure prayer and spiritual prayer. The difference between these that during pure prayer one's mind brims actively with stirrings and thoughts, while spiritual prayer does not entail any movement of the mind. Words give way to wonder. Spiritual prayer is the place of deep stillness:

> all kinds and habits of prayer which mankind prays unto God, have their term [terminate] in pure prayer. Lamentations and self-humiliations and beseechings and inner supplications and sweet tears and all other habits which prayer possesses . . . their boundary and the domain within which they are set in motion, is pure prayer. (1/XXII:112)

He does not want us to be content with where we are, but descend to the riskier depths of spiritual prayer. Paradoxically, he says this is, in a sense, not prayer at all, if we are used to defining prayer as the expression of our needs and hopes. We have left that behind:

> As soon as the spirit has crossed the boundary of pure prayer and proceeded onwards, there is neither prayer, nor emotions, nor tears . . . nor beseechings, nor desire, nor longing after any of those things which are hoped for in this world or in the world to be. Therefore there is no prayer beyond pure prayer, . . . but beyond this limit [the spirit] passes into awestruck wonder . . . (1/XXII:112)

---

1. See Mayes, *Diving for Pearls*. Extracts from Wensinck, *Mystical Treatises* are cited as, for example, as 1/ XX: 109. Extracts from the newly discovered *Second Part by Isaac of Nineveh* (trans. Brock) are cited as, for example, 2/ X: 32.

We are beckoned to leave behind words and prayers and sink deeper into a wordless communion with the Divine—a sense of being at one with God. Words are no longer needed. They become utterly redundant.

Isaac develops this through the imagery of descending through the levels of the ocean of grace. On the surface of life we experience turbulence with large waves crashing and foaming. Beneath the surface currents and streams of thought compete in the soul. But sink lower—as it were, go down diving towards the depths and it becomes strangely quiet and calm. The deeper we go in prayer the more silent it becomes. We allow ourselves to sink in the ocean of grace.

"The deepest depths of quiet in God." Isaac uses this phrase when reflecting on the three inter-linked realities of solitude, silence and stillness:

> When by prolonged solitude my heart has acquired peace from the trouble of recollections, solitude sends me continually waves of gladness which arise from emotions which burst forth from within unexpectedly and suddenly, to the delight of my heart; the which, running against the ship of my soul and withdrawing it from the sounds of the world [temptations] and from the life of the flesh, immerse it in the deepest depths of quiet in God. (1/LXVI:314)

Isaac recalls how Jesus himself models a pattern of withdrawal, detachment and stillness:

> The Savior . . . honored and loved stillness at all times, saying "Let us go to the wilderness to rest by ourselves" and "He sat down in a boat and went to a deserted region with his disciples." It was especially at these times that He drew Himself away from people and remained in stillness . . . For the instruction of the children of light who would travel afterwards in His footsteps following this new mode of life, He carried out this solitary converse with God. (2/XII:1)

We too can enjoy such encounters with God:

> A person who has stillness and the converse of knowledge will easily and quickly arrive at the love of God, and with the love of God he will draw close to perfect love of fellow human beings . . . we should leave the open space of struggles and give ourselves over to stillness. (2/X:33,37)

But Isaac is emphatic that it takes great discipline, both over the tongue and the fluttering mind, to reach this place of quietude:

There cannot be recollection of mind and purity in prayer without much vigilance over speech and action, as well as a guard over the senses; nor can the awareness that is given by grace come about unless a person has acquired much discernment by means of stillness. (2/XIV:10)

Let us rejoice, then, and give thanks to God that we have been held worthy, even for a small moment, to be able to escape from chatter and talk with the passions—for we are, albeit just for one moment, through converse with some excellent meditation... this cannot be acquired without the continual reading of Scripture in stillness and the reflective search for things hidden, and prayer. (2/XXIX:4)

One thing is sure, such silence is indispensable: "If someone does not have stillness, he will not come to know any one of these things [about God's love]" (2/XVIII:16). Isaac encourages us to "dive into the sea of stillness." (2/XXXIV:5) Are you ready, with Nathaniel and Isaac, to make that further descent? Let's go for it!

FIVE INVITATIONS AT THE START OF THIS RETREAT:

To plant a fig tree in your soul is to follow in Nathaniel's steps...

1. Seek out a Place of Stillness and Solitude beneath the Fig Tree

Discover a place and space of openness and reflection where you can be known by God and receive from him. Create an opportunity for undisturbed prayer and silence.

2. Open up a Space Where You can be True with God and be Yourself

Realize that God knows your heart's aches and longings right now.

Be honest with him about your questions and concerns as Nathaniel was with Philip "Can anything good come out of Nazareth?" Like Nathanial may we have no guile, let there be no concealment in our soul—don't use fig leaves to hide behind!

3. Receive the Lord's Affirmation

Jesus affirmed Nathanael in his identity—"An Israelite indeed . . . " What does the Lord say to you? He renews your baptismal affirmation: "You are my beloved son and daughter, with you I am well-pleased." What glimpses into your identity and vocation do you seek to receive in this retreat?

4. Be Open to the Divine Surprises

"You will see greater things than this. You will see heaven opened." You ain't seen nothing yet! Don't predict what will happen to you beneath the fig tree of this retreat. Do not rule anything out! As Nathaniel opens himself to Jesus, so Jesus will open himself to him.

5. Be Ready to Make Your Own Response to Jesus

For Nathaniel this was: "You are the King of Israel." What will it be for you? What will you have to say to Christ?

Dear reader, let me be a Philip to you in this retreat as you stand in the shoes of Nathaniel. Let me say to you "We have found him about whom Moses in the law and also the prophets wrote, Jesus son of Joseph from Nazareth. Come and see!"

PRAYER EXERCISE

Place yourself under the beckoning fig tree. Take deep breaths and relax. You have arrived. Take it all in. Catch your breath. As you ponder slowly the words of the psalm, ask yourself: "What am I looking for? What am I needing?" In the silence surrender such longings and this time of retreat to God.

> O LORD, you have searched me and known me.
> You know when I sit down and when I rise up;
> you discern my thoughts from far away.
> You search out my path and my lying down,
> and are acquainted with all my ways.
> Even before a word is on my tongue,
> O LORD, you know it completely.
> You hem me in, behind and before,

and lay your hand upon me.
Such knowledge is too wonderful for me;
it is so high that I cannot attain it . . .
For it was you who formed my inward parts;
you knit me together in my mother's womb.
I praise you, for I am fearfully and wonderfully made.
Wonderful are your works; that I know very well.
My frame was not hidden from you,
when I was being made in secret,
intricately woven in the depths of the earth.
Your eyes beheld my unformed substance.
In your book were written all the days
that were formed for me,
when none of them as yet existed.
How weighty to me are your thoughts,
O God! How vast is the sum of them!
I try to count them— they are more than the sand;
I come to the end— I am still with you. (Ps 139:1–6, 13–18)

VERDUN MEMORIAL OAK, LEOMINSTER

5   Planting an Oak

Celebrating the Journey

They will be called oaks of righteousness,
  the planting of the LORD, *to display his glory . . . (Isa 61:3)*

SIGNIFICANT MOMENTS IN LIFE'S journey are marked by the planting of trees.

We recalled the lad's planting of the conker in my garden in 1751. Near our house, in the grounds by Leominster Priory, a poignant oak tree was planted by Alderman Henry Gosling in 1921 from an acorn brought back from Verdun in France. The Battle of Verdun in France was fought over 303 days, the longest battle of the First World War, claiming three-quarter of a million casualties. It also destroyed 185,000 hectares of forest, mainly oak and chestnut, and the woodlands still bear the scars of the conflict. However an initiative led to the sending of acorns to England to be planted in remembrance of casualties. Nearby is found another, younger oak, planted in memory of those fallen in the Korean War 1950–53. But trees can also be planted to celebrate joyous occasions. South east of the Priory the community planted an apple orchard to mark the Millennium: any local resident is free to pick and use the fruit. Orchards indeed thrived around the Priory, and today a fine but solitary eating apple tree grows close the site where the original Benedictine High Altar once stood. Across town, on Ginhall Green, the community planted an orchard to mark the Queen's Diamond Jubilee in 2012. Such trees, planted in times of sadness and joy, can be a source of healing and comfort, and sustain memories of great moments in the life of communities.

THE OAKS OF MOREH AND MAMRE

In the Hebrew scriptures mighty oaks appear as landmarks on the journey, and as significant places to pause in one's journey and celebrate one's progress. They can help us reflect on our own life-pilgrimage.

In the book of Genesis the oaks of Moreh and Mamre feature at momentous events in the journeys of patriarchs Abraham and Jacob, and soar to the sky at the important stages on their itinerary through what today we call the West Bank at Shechem and Hebron.

1. Celebrate Your Enfolding Journey with Milestones

The Letter to Hebrews celebrates Abraham as the archetypal pilgrim:

> By faith Abraham obeyed when he was called to set out for a place that he was to receive as an inheritance; and by faith he stayed for a time in the land he had been promised, as in a foreign land, living in tents. For he looked forward to the city that has foundations, whose architect and builder is God. (Heb 11:8–10)

Genesis relates the episode where his journey from Haran in Mesopotamia towards Canaan began:

> Now the LORD said to Abram, 'Go from your country and your kindred and your father's house to the land that I will show you. I will make of you a great nation, and I will bless you, and make your name great, so that you will be a blessing. I will bless those who bless you, and the one who curses you I will curse; and in you all the families of the earth shall be blessed.'
>
> So Abram went, as the LORD had told him; and Lot went with him. Abram was seventy-five years old when he departed from Haran. Abram took his wife Sarai and his brother's son Lot, and all the possessions that they had gathered, and the persons whom they had acquired in Haran; and they set forth to go to the land of Canaan. When they had come to the land of Canaan, Abram passed through the land to the place at Shechem, to the oak of Moreh. At that time the Canaanites were in the land. Then the LORD appeared to Abram, and said, 'To your offspring I will give this land.' So he built there an altar to the LORD, who had appeared to him. From there he moved on to the hill country on the east of Bethel, and pitched his tent, with Bethel on the west and Ai on the east; and there he built an altar to the LORD and invoked the

5   PLANTING AN OAK

name of the LORD. And Abram journeyed on by stages towards the Negeb. (Gen 12:1–9)

Abram was prepared to move—physically and spiritually. God commands: "Go from your country." He was prepared to quit his comfort zone, leave behind his home, and venture forth on a journey into the unknown. He was prepared to let go of his familiar securities, even his cherished inherited concepts of God, and move out on a journey of faith: "he set out, not knowing where he was going" (Heb 11:8). Abraham truly had a pilgrim heart, or as the English poet of the 17th century from the Anglican tradition, George Herbert, put it: "a heart in pilgrimage." He was prepared to live with vulnerability and risk, quitting his comfortable stone house, ready to be "living in tents" (Heb 11:9).

Wearied by his journey, Abraham and Sarah, together with their caravan, see the mighty oak standing proud in the landscape as they pass through the valley between Mount Gerizim and Mount Ebal. It becomes a destination to aim for and turns out to be a place of significant encounter with the Divine. Abraham pauses at the Oak of Moreh and makes this feature in the landscape both a staging-post and a place of worship and sacrifice: "He built an altar there." The Oak of Moreh at Shechem in the Canaanite highlands marks the point of arrival in the promised land and it is beneath its bowers that he experiences some kind of theophany: "the LORD appeared to Abram." The tree marks a significant meeting with God and the first time God promises Abraham the gift of land for himself and his descendants.

We too find ourselves on a spiritual journey and we must keep moving. We have not arrived. We must be prepared, like Abraham himself, to let go sometimes of conventions and concepts that would pull us back and tie us down. We must be prepared, as it were, to "live in tents"—to live with the provisional, the impermanent, the uncomfortable, the unsettling, for as long as it takes. We do not know the outcome of our journey, but God does. God calls us to keep moving forwards. Let us pray for the faith of Abraham to grow in our own hearts, an unshakable trust in God, the God of our journey. Paul puts it: "you are Abraham's offspring, heirs according to the promise" (Gal 3:28).

We too can recall in our life a stage or situation that we longed for. We too can recall significant moments in our spiritual journey.

2. Pause, Worship and Reflect

According to the narrative in Genesis, Abraham crossed the desert of the Negev and journeyed on as far as Egypt because "there was a famine in the land" (12:10). Immediately, it seems, he enters a dry and challenging season in his pilgrimage. He does not rest on his laurels at Shechem and resumes his journey, but soon encounters trial and testing. But the journey is not over yet. Pharaoh casts him out and other mighty oaks beckon:

> So Abram went up from Egypt, he and his wife and all that he had, and Lot with him, into the Negeb. Now Abram was very rich in livestock, in silver, and in gold. He journeyed on by stages from the Negeb as far as Bethel, to the place where his tent had been at the beginning, between Bethel and Ai, to the place where he had made an altar at the first; and there Abram called on the LORD ... The LORD said to Abram, after Lot had separated from him, 'Raise your eyes now, and look from the place where you are, northwards and southwards and eastwards and westwards; for all the land that you see I will give to you and to your offspring for ever. I will make your offspring like the dust of the earth; so that if one can count the dust of the earth, your offspring also can be counted. Rise up, walk through the length and the breadth of the land, for I will give it to you.' So Abram moved his tent, and came and settled by the oaks of Mamre, which are at Hebron; and there he built an altar to the LORD. (Gen 13:1-4, 14-18)

Abraham finds himself in a land of conflict between warlords, according to Genesis chapter 14. He is settled for a season beneath the oak of Mamre but gets caught up into struggles of various kinds. But during this season of danger and turbulence, he encounters God in the darkness of the night:

> As the sun was going down, a deep sleep fell upon Abram, and a deep and terrifying darkness descended upon him. Then the LORD said to Abram, "Know this for certain, that your offspring shall be aliens in a land that is not theirs ... " When the sun had gone down and it was dark, a smoking fire pot and a flaming torch passed between these pieces [of sacrifice]. On that day the LORD made a covenant with Abram, saying, "To your descendants I give this land ... " (Gen 15: 12-13,17-18)

3. Stay Open to the Unexpected

The image of Abraham sitting at the entrance of his tent by the oaks of Mamre is a powerful symbol of expectancy and openness. What is Abraham dreaming about, longing for? Who might he expect?

> The LORD appeared to Abraham by the oaks of Mamre, as he sat at the entrance of his tent in the heat of the day. He looked up and saw three men standing near him. When he saw them, he ran from the tent entrance to meet them, and bowed down to the ground. He said, "My lord, if I find favor with you, do not pass by your servant. Let a little water be brought, and wash your feet, and rest yourselves under the tree. Let me bring a little bread, that you may refresh yourselves, and after that you may pass on—since you have come to your servant." So they said, "Do as you have said." And Abraham hastened into the tent to Sarah, and said, "Make ready quickly three measures of choice flour, knead it, and make cakes." Abraham ran to the herd, and took a calf, tender and good, and gave it to the servant, who hastened to prepare it. Then he took curds and milk and the calf that he had prepared, and set it before them; and he stood by them under the tree while they ate . . .
>
> The LORD said to Abraham, " . . . Is anything too wonderful for the LORD? At the set time I will return to you, in due season, and Sarah shall have a son." (Gen 18:1–8, 13–14)

Abraham exemplifies a capacity to welcome the stranger. For him, this turns out to be a life-changing encounter with the Divine. He is prepared to accept the stranger into his midst unconditionally, anticipating the call of God through the prophet Isaiah:

> Enlarge the site of your tent,
>   and let the curtains of your habitations be stretched out;
> do not hold back; lengthen your cords
>   and strengthen your stakes. (Isa 54:2)

Abraham emerges as open-hearted and prepared to receive astonishing news from the mysterious Angels. He is prepared, at the age of ninety nine, to accept a change in his vocation. The spreading and outreaching branches of the mighty oak seem to mirror the openness and acceptance of Abraham's soul. As The Letter to the Hebrews puts it:

> Do not neglect to show hospitality to strangers, for thereby some have entertained angels unawares. (13:2, RSV)

4. Celebrate Breakthroughs

While at Shechem and Hebron Abraham finds that existing trees become symbolic and significant, at Beer Sheba he decisively plants a new tree himself to mark and memorialize a breakthrough—a major step forward through the briars of mistrust and conflict.

In situations where strife could have occurred and then escalated, Abraham seems able to defuse them. He emerges as a peacemaker. His unselfish nature is not only seen in giving his nephew Lot first choice of land in which to pasture his flocks and herds (Gen 13:9). It reveals itself also in his readiness and determination to intercede for the condemned people of Sodom and Gomorrah (Gen 18:16–33). He did what he could to spare them from impending judgment.

Abraham's relationship with the Philistine warlord Abimelech of Gerar was at first, according to Genesis 20, fraught and full of suspicion and tension. In Genesis 21:25–34 we read of Abraham's argument with Abimelech, whose servants seized control of a well of water Abraham was using. But Abraham takes some creative steps forward to heal the fractured relationship. He resolves the dispute and avoids further conflict by entering into a covenant with Abimelech, sealing the agreement by his gift to him of seven lambs, which gives the well the name "Beersheba", meaning "the well of the oath." In this Abraham reveals a capacity for reconciliation and risk. His covenant or agreement with the Other is the first recorded in Scripture and encourages us to take risks that might lead to deeper mutual understanding and even friendship, deepening trust and openness:

> Therefore that place was called Beer-sheba; because there both of them swore an oath. When they had made a covenant at Beer-sheba, Abimelech, with Phicol the commander of his army, left and returned to the land of the Philistines. Abraham planted a tamarisk tree in Beer-sheba, and called there on the name of the LORD, the Everlasting God. And Abraham resided as an alien for many days in the land of the Philistines. (Gen 21:31–4)

Abraham plants this tree to stand as a memorial of the covenant he makes with one-time enemy Abimelech. For him, it marks the breaking down of barriers in a relationship, the overcoming of a stalemate, a real advance in building community and fostering peace. It prompts us to recall breakthroughs in our own lives: times when the impossible became possible, and small miracles of grace broke out—a step forward, or even a leap, in our spiritual journey...

The narrative tells us that "Abraham resided as an alien for many days in the land of the Philistines." He was prepared to abide for a significant amount of time in a foreign land as a sojourner, stranger, pilgrim. He was ready for the long haul, and was not going away—he displays a determination, a tenacity to remain in a place where he will encounter those of different traditions around him, in the hope that the Other will become Brother. Friend of God is prepared to become friend to all Others.

5. Grieve and Move On

> Sarah lived for one hundred and twenty-seven years; this was the length of Sarah's life. And Sarah died at Kiriath-arba (that is, Hebron) in the land of Canaan; and Abraham went in to mourn for Sarah and to weep for her. Abraham rose up from beside his dead, and said to the Hittites, "I am a stranger and an alien residing among you; give me property among you for a burying-place, so that I may bury my dead out of my sight." . . . So the field of Ephron in Machpelah, which was to the east of Mamre, the field with the cave that was in it and all the trees that were in the field, throughout its whole area, passed to Abraham as a possession in the presence of the Hittites, in the presence of all who went in at the gate of his city. After this, Abraham buried Sarah his wife in the cave of the field of Machpelah facing Mamre (that is, Hebron) in the land of Canaan. The field and the cave that is in it passed from the Hittites into Abraham's possession as a burying-place. (Gen 23:1–4, 17–20)

Near the oaks of Mamre Abraham buries his beloved wife. They had been through so much together and shared a long and eventful journey. Now is a time for letting go and leaving Sarah to rest forever beside the oaks of Mamre. But it is also a time for looking to the future. Genesis 24 tells us about the arrangements Abraham made to seek out a suitable wife for the child of promise Isaac—Rebeccah is found –while Abraham himself remarries and fathers six more children. Bereavement moves inexorably on to new beginnings.

THE OAKS OF MOREH REVISITED

If they could speak, the oaks of Moreh would have many more stories to tell. They continue to be significant throughout the unfolding story of the Israelites, especially in the experience of Jacob, Joshua and Jotham.

1. Jacob: Reject Idols

Throughout the story of the Israelites they are plagued by the temptation to idolatry. It is something that will not go away, and we have already noted how the prophets condemn the practice. Jacob attempts to deal decisively with this as God calls him away from Shechem, which had recently witnessed much bloodshed, and towards a new location:

> God said to Jacob, "Arise, go up to Bethel, and settle there. Make an altar there to the God who appeared to you when you fled from your brother Esau." So Jacob said to his household and to all who were with him, "Put away the foreign gods that are among you, and purify yourselves, and change your clothes; then come, let us go up to Bethel, that I may make an altar there to the God who answered me in the day of my distress and has been with me wherever I have gone." So they gave to Jacob all the foreign gods that they had, and the rings that were in their ears; and Jacob hid them under the oak that was near Shechem. (Gen 35:1–4)

The Oak of Moreh now becomes a place where wooden and metal idols would be interred, and other trinkets associated with the cult of idolatry. The very tree recalling Abraham's decisive break with the past now is a place for letting go of attachments that have become fruitless and soul-destroying.

2. Joshua: Reorientate, Make Decisions, Renew Covenant

Joshua, who had led the Israelites into the promised land in succession to Moses, has one great act to complete before he dies. He wants to give his people the opportunity to decisively renew their commitment to God and make an act of rededication. As Deuteronomy tells its story, this was somehow anticipated by Moses himself:

> See, I am setting before you today a blessing and a curse: the blessing, if you obey the commandments of the LORD your God that I am commanding you today; and the curse, if you do not obey the

5 PLANTING AN OAK

commandments of the LORD your God, but turn from the way that I am commanding you today, to follow other gods that you have not known. When the LORD your God has brought you into the land that you are entering to occupy, you shall set the blessing on Mount Gerizim and the curse on Mount Ebal. As you know, they are beyond the Jordan, some distance to the west, in the land of the Canaanites who live in the Arabah, opposite Gilgal, beside the oak of Moreh (Deut 11:26–30).

And so the time comes for this great act of renewal beneath the oaks of Moreh:

> Then Joshua gathered all the tribes of Israel to Shechem, and summoned the elders, the heads, the judges, and the officers of Israel; and they presented themselves before God. And Joshua said to all the people . . . "Choose this day whom you will serve, whether the gods your ancestors served in the region beyond the River or the gods of the Amorites in whose land you are living; but as for me and my household, we will serve the LORD."
>
> Then the people answered, "Far be it from us that we should forsake the LORD to serve other gods; for it is the LORD our God who brought us and our ancestors up from the land of Egypt, out of the house of slavery, and who did those great signs in our sight. He protected us along all the way that we went, and among all the peoples through whom we passed; and the LORD drove out before us all the peoples, the Amorites who lived in the land. Therefore we also will serve the LORD, for he is our God . . . The LORD our God we will serve, and him we will obey."
>
> So Joshua made a covenant with the people that day, and made statutes and ordinances for them at Shechem. Joshua wrote these words in the book of the law of God; and he took a large stone, and set it up there under the oak in the sanctuary of the LORD. Joshua said to all the people, "See, this stone shall be a witness against us; for it has heard all the words of the LORD that he spoke to us; therefore it shall be a witness against you, if you deal falsely with your God." So Joshua sent the people away to their inheritances. (Josh 24: 1, 15–18, 24–28)

It is still possible to stand of this very spot today in the West Bank outside Nablus at the site of the ancient city of Shechem located in the valley midway between mounts Ebal and Gerizim. The oak is long-gone but archeology has revealed at Tel Balata the walls of the temple of baal-berith—the temple of the Lord of the covenant. Indeed a stone akin to

Joshua's great monolith has been set up by this sanctuary, marking the place of the primordial oak. Poignantly, this site is close by Jacob's Well of John 4 which centuries later would come to witness similar decisiveness, dedication and commitment.

3. Jotham: Make Mistakes and Learn Afresh

At a turbulent period in Israel's history there were power struggles for the leadership of the people. Abimelech son of Gideon, put himself forward as King after the death of Gideon, but he was a violent man who had slaughtered most of his brothers, and was poor leader. Beneath the oak of Shechem the people made a big mistake when they crowned him at this spot: "Then all the lords of Shechem and all Beth-millo came together, and they went and made Abimelech king, by the oak of the pillar at Shechem" (Judg 9:6).

The narrative continues immediately with Abimelech's one surviving brother offering the parable of the trees, suggesting that this upstart candidate was a fearful bramble:

> When it was told to Jotham, he went and stood on the top of Mount Gerizim, and cried aloud and said to them, 'Listen to me, you lords of Shechem, so that God may listen to you.
> The trees once went out
>   to anoint a king over themselves.
> So they said to the olive tree,
>   "Reign over us."
> The olive tree answered them,
>   "Shall I stop producing my rich oil
>   by which gods and mortals are honored,
>   and go to sway over the trees?"
> Then the trees said to the fig tree,
>   "You come and reign over us."
> But the fig tree answered them,
>   "Shall I stop producing my sweetness
>   and my delicious fruit,
>   and go to sway over the trees?"
> Then the trees said to the vine,
>   "You come and reign over us."
> But the vine said to them,
>   "Shall I stop producing my wine
>   that cheers gods and mortals,
>   and go to sway over the trees?"

5 PLANTING AN OAK

So all the trees said to the bramble,
  "You come and reign over us."
And the bramble said to the trees,
  "If in good faith you are anointing me king over you,
    then come and take refuge in my shade;
  but if not, let fire come out of the bramble
    and devour the cedars of Lebanon."
'Now therefore, if you acted in good faith and honor when you made Abimelech king, . . . then rejoice in Abimelech, and let him also rejoice in you; but if not . . . let fire come out from the lords of Shechem, and from Beth-millo, and devour Abimelech.' (Judg 9: 7–16, 20).

This was the time of the Judges governing the people—the institution of the monarchy had not yet began, so anyone wanting to be king was acting presumptuously. In this, the earliest parable in the Scriptures, the olive, fig and vine all reply: "Shall we give up our calling, to bear valuable fruits for the good and enjoyment of God and mortals, and soar above the other trees?" This may represent humility and acceptance of vocation in contrast to the self-serving briar, ambitious and merciless. It is significant that Jotham offers this on well- forested Mount Gerizim soaring above Shechem and not on the bare-headed Ebal opposite. The very trees in the landscape suggest a language and an interpretation of events to Jotham.

THE PROGRESS OF A PILGRIM

As we reflect on the theme of spiritual journeys, we can be heartened by John Bunyan (1628–88). He wrote *Pilgrim's Progress* in 1678, reflecting something of his own eventful spiritual journey and celebrating the travels of the patriarchs:

> As I walked through the wilderness of the world, I came to a place where there was a den. There I lay down to sleep; and as I slept, I dreamed a dream. In my dream I saw a man clothed with rags, standing by a path with a book in his hand and a great burden upon his back . . . .
>
> Christian finally came to the little gate. Over the gate was written, in bold letters: "KNOCK, AND IT SHALL BE OPENED UNTO YOU." Christian knocked . . . At last one came to the gate whose name was Goodwill. He asked, in a deep voice, "Who's there, where did you come from, and what do you want?"

CHRISTIAN: I am a poor, burdened sinner. I come from the City of Destruction, and I want to go to Mount Zion, that I may be safe from the coming wrath of God. I am informed that through this gate is the way to Zion. I would like to know, therefore, if you will let me in.

GOODWILL . . . We do not reject any who come. No matter what they have done before coming, they are in no wise cast out. And now, my good pilgrim, come with me a little way, and I will show you the way to go. Now look yonder. Do you see that narrow way? That is the road you must take. It was travelled by the patriarchs in olden times, and by the prophets, and by Christ and His apostles; and it is as straight as a line can make it.

CHRISTIAN: But are there no turnings or windings by which a stranger may lose his way?

GOODWILL: Yes, there are many roads branching off from this one, but you can distinguish the right way from the wrong, for the right way is the only road that is straight and narrow . . .

At a great distance, Christian could see a magnificent mountainous country. In this faraway land were great forests, green vineyards, sparkling fountains, broad fields . . . He asked the name of the country. They said, "It is Immanuel's Land, and it is for all pilgrims, just as this hill is, and from there you will be able to see the gate of the celestial City, as the shepherds will show you."

He expressed his desire to go, and they were willing. "But first," they suggested, "let us go again to the storehouse." There they equipped him from head to foot with what he would need most on his journey. Being thus clothed, he walked out with his friends to the gate . . .

At the foot of the hill, Christian's good companions gave him a loaf of bread, a bottle of wine, and a large bunch of raisins. Bidding them good-bye, he went on alone.[1]

As we walk, with Christian, in the way of the patriarchs, we mark our progress with gratitude, and press on towards the great forests of Immanuel's land.

---

1. From Thomas, *Pilgrim's Progress in Today's English*.

5   PLANTING AN OAK

WHAT DOES IT MEAN TO PLANT AN OAK IN THE SOUL?

From Abraham's story we learned:

1. Mark your enfolding journey with milestones
2. Pause, worship and reflect
3. Stay open to the unexpected
4. Celebrate breakthroughs
5. Let go, grieve and move on

From the oaks of Moreh at Shechem we see three imperatives:

1. Jacob: reject idols
2. Joshua: reorientate, make decisions, renew covenant
3. Jotham: make mistakes and learn afresh

To plant an oak in the soul suggests to us these significant and memorable moments on our own pilgrimage of faith:

- We can celebrate and give thanks for landmarks in our spiritual journey and look back to see where we have come from over the years.
- We can be thankful for "settlements" in a life of transition and change. There are times and seasons in the spiritual life. A time to settle and know some stability is vital.
- We can open ourselves up to the other and to the stranger, and to God's surprising blessings (as in gift of Isaac).
- We can allow ourselves to take risks in friendship and risks in reconciliation.
- We can permit ourselves to grieve over bereavements, whether the loss of a loved-one or friend or relationship, or the loss of an opportunity or job.
- We can let go of things that formerly claimed our attention.

- We can decisively renew our relationship with God and make recommitments.
- We can allow ourselves to make mistakes from time to time.

PRAYER EXERCISE

On a clean piece of paper (landscape) draw a personal "timeline" to recall the transitions you have faced in your life.

Draw a horizontal line and mark it into the decades of your life.

Above the line, the space representing the visible world, plant oaks—in a simple line drawing—and label, noting major events and transitions, including new jobs, house-moves, births and deaths, new ministries. Plant an oak, as it were, in remembrance, thanksgiving, celebration, penitence, marking it on the timeline.

Below the line, in the hidden underground section try to note how you felt at these moments of change. How did you experience God at these moments? Maybe trace roots from the base of the tree that you can label with appropriate words recalling your hidden discoveries at that time. Notice, too, how the roots may be intertwined where one transition has affected another.

Be guided by the occasions for planting oaks we noticed in the book of Genesis.

If in a group setting, you might like to reflect on this with a partner.

Bring this to a close by giving thanks for God's providence in your life, and entrust your future to him, ending with this act of dedication and affirmation:

> The Spirit of the Lord GOD is upon me,
>   because the LORD has anointed me;
> he has sent me to bring good news to the oppressed,
>   to bind up the brokenhearted,
> to proclaim liberty to the captives,
>   and release to the prisoners;
> to proclaim the year of the LORD's favor . . .
>   to provide for those who mourn in Zion—

5 PLANTING AN OAK

> to give them a garland instead of ashes,
> the oil of gladness instead of mourning,
>   the mantle of praise instead of a faint spirit.
> They will be called oaks of righteousness,
>   the planting of the Lord, to display his glory . . .
> I will greatly rejoice in the Lord,
>   my whole being shall exult in my God . . .
> For as the earth brings forth its shoots,
>   and as a garden causes what is sown in it to spring up,
> so the Lord God will cause righteousness and praise
>   to spring up before all the nations. (Isa 61:1–3, 10–11)

HARPS HUNG ON WILLOW

6   Planting a Willow

Letting Go, Moving On

"Go forward in procession with branches" (Ps 118:27)

TWO TYPES OF WILLOW flourish in our gardens at Borderlands Retreats. A mighty goat willow, *Salix caprea* is a hundred years old, its bark grey-brown marked with diamond-shaped fissures. This is the pussy willow—a beautiful harbinger of spring as the catkins of late winter emerge, leading to a great canopy of oval leaves on drooping branches in summer. This evokes the pain and anguish of Psalm 137. Goat willow was known as Palm Willow and was commonly cut for Palm Sunday celebrations and carried to church. The Palm Willow was then kept through the year to protect the home against thunder, lightning, disease and other dangers.

We also have thin willows *Salix vimnalis* forming a graceful fence, its replanted off-cuts producing new unexpected growths—evoking the joyful procession into the future expressed in Psalm 118 and in the Jewish Liturgy of the Feast of Tabernacles. While the first willow speaks of letting go of hurts and attachments, the second speaks of moving decisively into God's future. In this session we will explore these two movements of the soul.

1. WILLOWS OF PAIN AND RELEASE

> By the rivers of Babylon—
>   there we sat down and there we wept
>   when we remembered Zion.
> For there our captors
>   asked us for songs,
> and our tormentors asked for mirth, saying,

> "Sing us one of the songs of Zion!"
>> How could we sing the LORD's song
>>> in a foreign land?
>> If I forget you, O Jerusalem,
>>> let my right hand wither!
>> Let my tongue cling to the roof of my mouth,
>>> if I do not remember you,
>> if I do not set Jerusalem
>>> above my highest joy. (Ps 137)

One can feel the pain and heartache in this poignant psalm, first uttered, it seems, by the willows overhanging a river in the land of exile. What do the willows teach us? What does it mean to plant a willow in the soul? In this psalm, lamentation turns to resolution. The identification and naming of pain, and giving this expression, moves towards renewed commitments.

What Do the Harps Represent?

On the willows there we hung up our harps.

From the dawn of time the harp has been a poignant symbol of freedom, self-expression and joyous worship. Genesis relates that seven generations after Adam "Lamech and Adah had two sons, Jabal and Jubal. Their son Jabal was the first to live in tents and raise sheep and goats. Jubal was the first to play harps and flutes" (4:21, CEV). The scripture tells us that the harp was the chief instrument of worship:

> David and all Israel were dancing before God with all their might, with song and lyres and harps and tambourines and cymbals and trumpets. (1 Chr 13:8)

> David also commanded the chiefs of the Levites to appoint their kindred as the singers to play on musical instruments, on harps and lyres and cymbals, to raise loud sounds of joy. (1 Chr 15:16)

> Then I will go to the altar of God, to God my exceeding joy; and I will praise you with the harp, O God, my God. (Ps 43:4)

> I will incline my ear to a proverb; I will solve my riddle to the music of the harp. (Ps 49:4)

> Awake, my soul! Awake, O harp and lyre! I will awake the dawn. (Ps 57:8)

The harp represents the way people used to pray and worship before the disastrous event of 586BC:

> The king of the Chaldeans killed their youths with the sword in the house of their sanctuary, and had no compassion on young man or young woman, the aged or the feeble; he gave them all into his hand. All the vessels of the house of God, large and small, and the treasures of the house of the Lord, and the treasures of the king and of his officials, all these he brought to Babylon. They burned the house of God, broke down the wall of Jerusalem, burned all its palaces with fire, and destroyed all its precious vessels. He took into exile in Babylon those who had escaped from the sword, and they became servants to him and to his sons until the establishment of the kingdom of Persia, to fulfil the word of the Lord by the mouth of Jeremiah, until the land had made up for its sabbaths. All the days that it lay desolate it kept sabbath, to fulfil seventy years. (2 Chr 36:17–21)

Within the experience of physical exile and spiritual displacement, when the Jerusalemites had become refugees in an alien land, the harp took on a deeper, poignant meaning. It now represented the way things were in the past, the songs we used to sing, the way we used to worship in the temple. It became a symbol of the past, of a world that was lost, ripped away from the people. As the now faraway temple lay in ruins, its treasures stolen and carried off as booty, the harp represents something that has been interrupted, left incomplete, unfulfilled, something unfinished. It stands for something left hanging, broken dreams, love unrequited. Now it was time, by the rivers of Babylon, to do four things:

We sat down. We need time to process what has gone on. We need a space for healing.

We wept. We need to lament, grieve, shed tears maybe for what has been lost, for what remains unfulfilled, for broken dreams and shattered hopes. The Weeping Willow is so-called because tears of rain drip down its branches: it becomes a powerful image of the sorrowful soul.[1]

We remembered. We remembered Zion. We recall the world we have left behind, friends and family from whom we are parted by bereavement or dislocation. We recall patterns of prayer and worship we once enjoyed. We

---

1. The weeping willow tree is known as *Salix babylonica*.

think of the security of the traditional and familiar which has been taken away from us or which we have had to leave behind.

We hung up our harps. On the willows we hang our harps not in resignation or in anger, but in sorrow mingled with hope. There was a letting go. This was strangely a healing moment, a release. It filled us with new hope and resolution. Now we could express these commitments:

I will never forget you, O Jerusalem,
I will always set Jerusalem above my highest joy

In lonely exile here?

When in 1861 John Mason Neale wrote his hymn based on the eleventh century Antiphons for late Advent, he expressed humanity's longing for a redeemer in the imagery of a present-day exile, akin to that of old. The experience of exile is ours:

> O come, O come, Emmanuel,
> and ransom captive Israel,
> that mourns in lonely exile here,
> until the Son of God appear.
> Rejoice! Rejoice! Emmanuel
> shall come to thee, O Israel.

More recently theologians Stanley Hauerwas and William Willimon in their book *Resident Aliens* noticed that this kind of metaphor is very pertinent to the counter-cultural situation of the church in today's world, finding itself displaced from center-stage and now a voice in the margins:

> The Jews in Dispersion were well acquainted with what it meant to live as strangers in a strange land, aliens trying to stake out a living on someone else's turf . . . We believe that the designation of the church as a colony and Christians as resident aliens are not too strong for the modern church—indeed, we believe it is the nature of the church, at any time and in any situation, to be a colony . . . an island of one culture in the middle of another.[2]

If the experience and language of exile resonates with the precarious position of the church in today's postmodern world, so too does it ring bells with our experience of the pandemic. This time of international crisis

---

2. Hauerwas and Willimon, *Resident Aliens*, 11,12.

impacted us personally, and many have spoken in terms of an inner displacement, a loss of familiar routines—we found ourselves in a new place, and we didn't like it very much. Landmarks that we relied on regularly crumbled. We were dislocated spiritually, out of touch with the familiar routines of life, and for those prevented from going to church or other social activities, it was perceived as a "lonely exile here."

The ancient tear-filled psalm finds an echo in our soul. We glimpse new meanings in exile, harp and willow. Fundamentally, the message is the same from the sixth century BC to today's situation. We are called to a radical letting go, and though painful, this becomes a release and a liberation.

The First Willow Calls us to the Prayer of Relinquishment

This is a prayer of radical surrender to God, made not in resignation but in trust. It is akin to Christ's *kenosis* as expressed in Philippians 2: "he did not cling . . . he emptied himself." Christ lays down his deity in order to become a servant of all. The costly release is for a good purpose and for the sake of serving others. Sometimes what we relinquish needs to die so that God can accomplish his purposes through us: "I have been crucified with Christ; and it is no longer I who live, but it is Christ who lives in me. And the life I now live in the flesh I live by faith in the Son of God, who loved me and gave himself for me" (Gal 2:19b-20). This is a fundamental letting-go, in which our resistances to God crumble away, and self-will and ambition yield to an echo of the Gethsemane prayer: "Abba, Father . . . Not my will, but yours be done."

It is expressed in the Methodist Covenant Prayer:

> I am no longer my own but yours.
> Put me to what you will,
> rank me with whom you will;
> put me to doing,
> put me to suffering;
> let me be employed for you,
> or laid aside for you,
> exalted for you,
> or brought low for you;
> let me be full,
> let me be empty,
> let me have all things,
> let me have nothing:

I freely and wholeheartedly yield all things
to your pleasure and disposal.
And now, glorious and blessed God,
Father, Son and Holy Spirit,
you are mine and I am yours. So be it.
And the covenant now made on earth, let it be ratified in heaven.

But the harp we hang up in surrender and letting go can represent whatever we need to release our grip from.

What harp do you need to hang up?

1. It may be something that belongs to the past—a nostalgia of how things used to be or a cherished tradition past its "sell-by" date.
2. It may be an old way of praying or worshipping or religious habit that does not quite work in the present situation—as the harps were not quite right for the situation in exile.
3. It may be a cloying obsession, a clamoring addiction, a routine you are overly-attached to but doesn't fit with circumstances today, or "a sin that clings so closely" (Heb 12:1).
4. It may be an attachment which is starting to dominate, physical or symbolic, or indeed a relationship that has become less than life-giving.
5. It may be an anger, a disappointment or frustration.
6. It may be an unforgiveness, a root of bitterness that can be dissolved by a courageous, God-empowered act of unilateral forgiveness.
7. It may be a desire to always be in control, which might represent a lack of joyous trust in God and his unfolding purposes.

We pray for self-knowledge and discernment as we seek to identify and give name to our harp.

We face the need for lament and griefwork, for tears and sighs.

The letting-go may be at the same time costly yet healing.

We pray for the gift of decisiveness and determination as with God's grace we lay down what perhaps we have been carrying for years, be it a burden, habit or attachment.

What Do You Need to Embrace?

Standing beneath the willow, we are called to fresh abandonment to God, a yielding, a submission, a giving in, a movement from hesitant holding back to courageous self-emptying.

In his great classic *Self-Abandonment to Divine Providence*, Jean-Pierre de Caussade (1675–1751) encourages us to abide in a state of surrender to God.[3] De Caussade urged his readers to strive for a synergy, an active cooperation with God's will: "We know that in all things God works for good for those who love him, who are called according to his purpose" (Rom 8:28, alt. reading). De Caussade believed that God is supremely active in the world, guiding all things according to his divine plans. Our part is to be awake and responsive to God's actions, to allow him to move and direct our life in the midst of change. We are to train ourselves to recognize God's hand of providence in the "chances and changes of this mortal life."

De Caussade gave us the striking phrase "the sacrament of the present moment." He teaches us that we should not live in the past nor become anxious about the future, but rather be totally available to God this day and this very moment: "See, now is the acceptable time; see, now is the day of salvation" (2 Cor 6:2). Today, right now, God waits to meet us. De Caussade urges us to live in an attitude of continual surrender to God, yielding ourselves totally to him without qualification or preconditions, so we can become channels through which he can work: "Loving, we wish to be the instrument of his action so that his love can operate in and through us."[4] We are to live by humble trust in God, confident that he is working his purposes out. We are not to seek our own fulfilment but God's Kingdom: "Follow your path without a map, not knowing the way, and all will be revealed to you. Seek only God's Kingdom and his justice through love and obedience, and all will be granted to you."[5] Abandoned into God's hands, we are to "go with the flow" as he opens and closes doors before us.

But what if our prayer resembles Gethsemane, and suffering and upheavals come our way—can these be welcomed as God's will for us? Should

---

3. Muggeridge, *Sacrament*. De Caussade sought to counter the heresy of Quietism, which taught that the surest way to union with God was to foster a state of utter passivity before God, necessitating a complete withdrawal from the world, the annihilation of the human will, and a cessation of all human effort, in the search to become totally available to God.

4. Muggeridge, *Sacrament*, 46.

5. Muggeridge, *Sacrament*, 75.

we not try to fight against them? De Caussade warns that we must not set bounds or limits to God's plans. He is a "God of surprises." As we saw in the experience of the exile, he works in unpredictable and unlikely ways and we should be ready for anything: "The terrifying objects put in our way are nothing. They are only summoned to embellish our lives with glorious adventures."[6] Hardships can be in God's hands pathways to growth: "With God, the more we seem to loose, the more we gain. The more he takes from us materially, the more he gives spiritually."[7] We should not resent difficult circumstances, but rather listen to what God is saying to us through them.

How then is it possible to cultivate an attitude of such openness to God? De Caussade affirms that it is achieved by living in communion with God, and allowing Jesus Christ to dwell at the very center of our being. The Christ who longs to live within us is "noble, loving, free, serene, and fearless."[8] De Caussade has a vision of the Christ-life growing within each person who has the courage to surrender to him. This is the secret of welcoming the fruit of "the sacrament of the present moment":

The mysterious growth of Jesus Christ in our heart is the accomplishment of God's purpose, the fruit of his grace and divine will. This fruit forms, grows and ripens in the succession of our duties to the present which are continually being replenished by God, so that obeying them is always the best we can do. We must offer no resistance and abandon ourselves to his divine will in perfect trust.[9]

2. WILLOWS OF CELEBRATION AND HOPE

Isaiah uses the willow image to speak of renewal and return to Zion—life in the post exilic era:

> But now hear, O Jacob my servant,
>   Israel whom I have chosen!
> Thus says the LORD who made you,
>   who formed you in the womb and will help you:
> Do not fear, O Jacob my servant,
>   Jeshurun whom I have chosen.
> For I will pour water on the thirsty land,

---

6. Muggeridge, *Sacrament*, 40.
7. Muggeridge, *Sacrament*, 54.
8. Muggeridge, *Sacrament*, 109.
9. Muggeridge, *Sacrament*, 111.

> and streams on the dry ground;
> I will pour my spirit upon your descendants,
>   and my blessing on your offspring.
> They shall spring up like a green tamarisk,
>   like willows by flowing streams. (Isa 44:1–4)

In 516 BC the exiles returned to Jerusalem from Babylon, and the temple was rebuilt. The festivals could resume and develop, now colored both by the painful experience of exile and the joyful new beginning of the return. The feast of Tabernacles (Succoth), originally a harvest festival, also celebrated the Exodus and asked the worshippers to remember their vocation to be pilgrims. During the week-long festival people lived in shelters built using the branches of the willow, as a lived-out reminder of their pilgrim status:

> On the first day you shall take the fruit of majestic trees, branches of palm trees, boughs of leafy trees, and willows of the brook; and you shall rejoice before the LORD your God for seven days. You shall keep it as a festival to the LORD seven days in the year; you shall keep it in the seventh month as a statute forever throughout your generations. You shall live in booths for seven days; all that are citizens in Israel shall live in booths, so that your generations may know that I made the people of Israel live in booths when I brought them out of the land of Egypt: I am the LORD your God. (Lev 23:40–43)

Willows were not only used for the construction of tabernacles, but also played a key role in liturgical processions: "Go forward in procession with branches, even to the altar!" (Ps 118:25)

People would gather up branches of willows, approach the temple, and lean them against the sides of the altar in front of the temple, their tops bent against the edges of the altar. As the shofar sounded each day, they would make a circuit around the altar once and recite the verse "Please, LORD, save us; please, LORD, give us success!" (Ps 118:25). On Hoshana Rabbah, the final day of Sukkoth, they made a circuit around the altar seven times. (These ceremonies are still enacted in the local synagogue). This may evoke the seven encirclements of the city of Jericho and the falling down of the city walls under Joshua's leadership—celebrating walls and barriers falling by the power of prayer. This prayer is repeated:

> Please, God please, save please and deliver please, you are our Father!

> Open the gates of heaven and open up the storage rooms of your bounty to us!
> You will save us and not extend the quarrel; deliver us, God our savior![10]

A final enigmatic prayer in the ceremony is said as people hold sprigs of willow in their hands. The refrain reads: "A voice is speaking, a voice is announcing, and says . . . " A pregnant pause follows. This refers to the voice of God guiding his people, and no words are supplied in the liturgy. The worshippers are kept hanging on in expectancy, called to stay listening out for the voice of God in their daily lives.

The painful poignant Psalm 137 is now displaced by the thankful celebratory Psalm 118:

> Thank the LORD because he is good.
>   His love continues forever.
> Let the people of Israel say,
>   "His love continues forever." . . .
> I was in trouble, so I called to the LORD.
>   The LORD answered me and set me free.
> I will not be afraid, because the LORD is with me.
>   People can't do anything to me . . .
> The LORD gives me strength and a song.
>   He has saved me.
> Shouts of joy and victory
>   come from the tents of those who do right:
>   "The LORD has done powerful things."
> The power of the LORD has won the victory;
>   with his power the LORD has done mighty things.
>   I will not die, but live,
>   and I will tell what the LORD has done.
>   The LORD has taught me a hard lesson,
>   but he did not let me die.
> Open for me the Temple gates.
>   Then I will come in and thank the LORD . . .
> This is the day that the LORD has made.
>   Let us rejoice and be glad today!
>   Please, LORD, save us;
>   please, LORD, give us success.
>   God bless the one who comes in the name of the LORD.
>   We bless all of you from the Temple of the LORD.
>   The LORD is God,

---

10. Farber, "Mystical Ritual."

and he has shown kindness to us.
With branches in your hands, join the feast.
Come to the corners of the altar.
You are my God, and I will thank you;
you are my God, and I will praise your greatness.
Thank the LORD because he is good.
His love continues forever. (Ps 118:1–2, 5–6, 14–19, 24–29 NCV)

TO PLANT A WILLOW IN THE SOUL IS TO MAKE 2 COMMITMENTS

"We hung up our harps"—we create a space in our lives when we can face up to attachments and decisively let go of what is not now needed.

"Go forward in procession with branches"—having given ourself a chance to take stock, we take time to celebrate God's healing grace and providence in our lives, and step into the future with new confidence.

Willows that once sorrowfully bore the harps in exile are now waved in exultation and hope.

PRAYER EXERCISE

This exercise is in 2 corresponding parts

First Read Psalm 137 verses 1–6.

Re-read the passage above "What harp do you need to hang up?" Praying for self-knowledge, give name to your harp. It will not be wrested from you—you will freely part from it.

On the retreat at Borderlands we make a simple harp from balsa wood.[11] You can draw one simply in a triangular D shape with pointed corners. On the side of the harp write down what you want this harp to represent or draw a suitable symbol.

Approach the willow (or a cross) in a spirit of trusting surrender to God. Do not hold back. When you are ready hand the harp over a protruding

---

11. Using the Woodcraft Construction Kit.

branch. You have done it! Step back and see it hanging there. You are clinging to it no longer and it has ceased calling out for attention and trying to control you. With tears or laughter renew your decisiveness. What do you feel?

Secondly read Psalm 118.

As you read it celebrate God's providence in your life. Call to mind the times you have experienced his guidance and protection. Choose one line in the psalm to repeat to yourself as you walk from the willow into the future. And give thanks.

BENEATH THE VINE

7 Planting a Vine

Abiding, Fruiting

> *They shall beat their swords into plowshares,*
>   *and their spears into pruning hooks;*
> *nation shall not lift up sword against nation,*
>   *neither shall they learn war any more;*
> *but they shall all sit under their own vines and under their own fig trees,*
>   *and no one shall make them afraid;*
>   *for the mouth of the* LORD *of hosts has spoken. (Mic 4:3,4)*
> *I will remove the guilt of this land in a single day. On that day, says the* LORD *of hosts, you shall invite each other to come under your vine and fig tree." (Zech 3:9,10)*

OUR HOME IN HEREFORDSHIRE is called Vine Cottage, and vines spread across the house and garden wall. As I sit beneath the vine I recall these two bible passages that celebrate sitting beneath the vine as a symbol of finding peace and God-given contentment—indeed, for Zechariah this is a symbol of the longed-for messianic age. A central image in the Last Supper discourses as John's Gospel gives them to us, it challenges us with a powerful paradox—we are to abide and we are to go.

ABIDE

> Abide in me as I abide in you . . . Those who abide in me and I in them bear much fruit, because apart from me you can do nothing . . . As the Father has loved me, so I have loved you; abide in my love. If you keep my commandments, you will abide in my love,

7 PLANTING A VINE

just as I have kept my Father's commandments and abide in his love. I have said these things to you so that my joy may be in you, and that your joy may be complete. (John 15:4,5, 9–11)

Other translations cast light on the meaning of this metaphor:

I am the real vine, my Father is the vine-dresser. He removes any of my branches which are not bearing fruit and he prunes every branch that does bear fruit to increase its yield. Now, you have already been pruned by my words. You must go on growing in me and I will grow in you. For just as the branch cannot bear any fruit unless it shares the life of the vine, so you can produce nothing unless you go on growing in me. I am the vine itself, you are the branches. It is the one who shares my life and whose life I share who proves fruitful. For the plain fact is that apart from me you can do nothing at all. The one who does not share my life is like a branch that is broken off and withers away. He becomes just like the dry sticks that men pick up and use for the firewood. But if you live your life in me, and my words live in your hearts, you can ask for whatever you like and it will come true for you. This is how my Father will be glorified—in your becoming fruitful and being my disciples. (*Phillips*)

I am the Vine, you are the branches. When you're joined with me and I with you, the relation intimate and organic, the harvest is sure to be abundant. Separated, you can't produce a thing. Anyone who separates from me is deadwood, gathered up and thrown on the bonfire. But if you make yourselves at home with me and my words are at home in you, you can be sure that whatever you ask will be listened to and acted upon. This is how my Father shows who he is—when you produce grapes, when you mature as my disciples. (*Message*)

I am the sprouting vine and you're my branches. As you live in union with me as your source, fruitfulness will stream from within you—but when you live separated from me you are powerless. (*Passion* translation)

John uses the word "abide" no less than eleven times in chapter 15. The Greek word *meno* which we translate as "abide" has a rich range of meanings. It can be translated by these words—which become imperatives from the mouth of Christ—stay, wait, linger, remain, stand your ground! When you find yourself in a place of stillness, stay at it, don't wriggle—allow yourself to be held by God in the stillness. Enjoy and relish the communion with

God. Savor his love lavished upon you (1 John 3:1). In this hyper-active and frantic world, just *abide*. Enjoy a sense of the abode—for this is in some sense a home-coming. Richard Foster calls prayer "finding the heart's true home." He writes:

> For too long we have been in a far country: a country of noise and hurry and crowds, a country of climb and push and shove, a country of frustration and fear and intimidation. God welcomes us home: home to serenity and peace and joy, home to friendship and fellowship and openness, home to intimacy and acceptance and affirmation.[1]

GO AND BEAR FRUIT

> This is my commandment, that you love one another as I have loved you. No one has greater love than this, to lay down one's life for one's friends.
> You did not choose me but I chose you. And I appointed you to go and bear fruit, fruit that will last . . .
> When the Counselor comes, whom I shall send to you from the Father, even the Spirit of truth, who proceeds from the Father, he will bear witness to me; and you also are witnesses, because you have been with me from the beginning. (John 15: 12–13, 16, 26–27)

As John gives us this discourse, in the next breath, having given the imperative to abide and stay still, Jesus goes on to command his disciples to move—to go and bear fruit. The disciples are to be *marturia*. The word, of course means witnesses, sharing one's first-hand experience of Christ with others, giving testimony, but it also gives rise to the word *martyrs*. Jesus is quite clear:

> Love each other *as* I have loved you. This is what I'm commanding you to do. The greatest love you can show is to give your life for your friends. You are my friends if you obey my commandments. (John 15: 12–14, *Names of God* translation)

That little word "as" (Greek *kathos*) means "like" or "in the same way": "Love one another the way I loved you. This is the very best way to love. Put your life on the line for your friends" (John 15:13, *Message*). We are being

---

1. Foster, *Prayer*, 1.

7 PLANTING A VINE

called to a Christ-like life: not only to abide in Christ as Christ abides in his Father (15:10) but, like him, to live sacrificially for others.

The experience of abiding, communion, is but a preparation for self-sacrifice. Grapes not to be admired but crushed, to release their goodness or else they wither on the vine. As Christ lingered in Gethsemane, the place of the oil press, so the wine press is inescapable. As Ephrem puts it, referring to Christ the Grape:

> The Grape was pressed and gave
> the Medicine of Life to the Nations.[2]

We are not to luxuriate selfishly in a passive quietude. We are to position ourselves in a place where we can receive and drink in the love of Christ which energizes us for courageous and risky ministry. We draw up the goodness and nutrients we need as we abide, but there must be a readiness—in the biblical imagery—for the blood of grapes to be spilt:

> Remember the days of old,
>   consider the years long past;
> ask your father, and he will inform you;
>   your elders, and they will tell you.
> The LORD's own portion was his people,
>   Jacob his allotted share.
> The LORD alone guided him;
>   no foreign god was with him.
> He set him atop the heights of the land,
>   and fed him with produce of the field . . .
>   you drank fine wine from the blood of grapes (Deut 32: 7, 9, 13, 14)

> Now I was the last to keep vigil;
>   I was like a gleaner following the grape-pickers;
> by the blessing of the Lord I arrived first,
>   and like a grape-picker I filled my wine press.
> Consider that I have not labored for myself alone,
>   but for all who seek instruction.
> Hear me, you who are great among the people,
>   and you leaders of the congregation, pay heed! (Sir 33:17–19)

> Finishing the service at the altars
>   and arranging the offering to the Most High, the Almighty,
> he [the high priest] held out his hand for the cup

---

2. "Hymns on Virginity 31:13" in Murray, *Symbols of Church and Kingdom*, 120.

and poured a drink offering of the blood of the grape;
he poured it out at the foot of the altar,
a pleasing odor to the Most High, the king of all. (Sir 50:14,15)

We recall the dramatic words which Christians came to associate with the passion, and recall each Good Friday:

"Why are your robes so red,
and your garments like theirs who tread the wine press?"
"I have trodden the wine press alone,
and from the peoples no one was with me;
I trod them in my anger
and trampled them in my wrath;
their juice spattered on my garments,
and stained all my robes.
For the day of vengeance was in my heart,
and the year for my redeeming work had come. (Isa 63:2–4)

Abiding in the vine renews the love of Christ in us, that we may bear a fruit that needs to be crushed and squeezed before it can be received. Love is measured by sacrifice. As we consider this, we do not forget that Christ promises us that the Counsellor, the Advocate, the Holy Spirit will always be with us (John 15:26).

FOUR SAINTS SHARE WITH US THIS INTERPLAY BETWEEN CONTEMPLATION AND ACTION.

Francis' Dilemma

Even St Francis (1181–1226) wrestled over the question of the relationship between activity and stillness and found himself torn between the two, as Bonaventure relates in his biography. Francis agonizes:

> What do you think, brothers, what do you judge better? That I should spend my time in prayer, or that I should travel about preaching? . . . in prayer there seems to be a profit and an accumulation of graces, but in preaching a distribution of gifts already given from heaven.

He went on to rehearse the advantages of a life dedicated solely to prayer:

7 PLANTING A VINE

> In prayer there is a purification of interior affections and a uniting to the one, true and supreme good with an invigorating of virtue; in preaching, there is dust on our spiritual feet, distraction over many things and a relaxation of discipline.

Ultimately he sees the truth that proves to be decisive:

> There is one thing ... that seems to outweigh all these considerations before God, that is, the only begotten Son of God, who is the highest wisdom, came down from the bosom of the Father for the salvation of souls in order to instruct the world by his example and to speak the word of salvation to people ... holding back for himself absolutely nothing that he could freely give for our salvation. And because we should do everything according to the pattern shown us in him ... it seems more pleasing to God that I interrupt my quiet and go out to labor.[3]

According to *The Little Flowers of St Francis* (ch. 16) Sister Clare and Brother Silvester, after a time of prayer, agree on the same advice to Francis: "Continue with your preaching, because God called you not for your sake alone but for the salvation of others".[4] However, in the life of Francis, this was never going to be an "either/or" choice. In the course of his mission, he established hermitages and retreats, and his whole ministry was an ebb and flow of action and contemplation. He models an integration of prayer into service, an inter-penetration and cross-fertilization between the two. It has been written of Francis: "his mystical experience, far from cutting him off from the world, always sent him right back into its most basic realities."[5] The presence of God is not only to be found in stillness and solitude, but, as Francis would put it, in the despised leper and the feared wolf.

Francis composed *A Rule for Hermitages* showing that he values solitude amidst activity. In this text Francis puts only one biblical text. It related to the Kingdom: "And let them seek first of all the Kingdom of God and his justice" (Matt 6:33). Thomas Merton observes:

> The importance of the document lies in the spirit which it exhales- a spirit of simplicity and charity which pervades even the life of solitary contemplation. It has been noted that the genius of sanctity is notable for the way in which it easily reconciles what

---

3. Bonaventure's "Major Legend of Saint Francis, ch.12" in Armstrong et al. (eds.), *Francis of Assisi Early Documents: Vol.* 2, 622.
4. Blaiklock and Keys (trans.), *Little Flowers*, 54.
5. Rotzetter et al., *Gospel Living*, 180.

seems at first sight irreconcilable. Here St Francis has completely reconciled the life of solitary prayer with warm and open fraternal love. Instead of detailing the austerities and penances which hermits must perform, the hours they must devote to prayer and so on, the saint simply communicates the atmosphere of love which is to form the ideal climate of prayer in the hermitage. The spirit of the eremitical life as seen by St Francis is therefore cleansed of any taint of selfishness and individualism. Solitude is surrounded by fraternal care and is therefore solidly established in the life of the Order and of the Church. It is not an individualistic exploit in which the hermit by the power of his own asceticism gains a right to isolation from an elevation above others.[6]

St Francis' life of witness culminated in the experience of receiving the stigmata on Mt Alverna: the very wounds of Christ appeared in his own feet, hands and side. But this was not a private ecstasy. Rather, Bonaventure tells us in his *Major Legend*, it led Francis to fresh engagement with the lepers: "Let us begin, brothers, to serve the Lord our God, for up to now we have done little." Bonaventure tells us: "He burned with a great desire to return to the humility he preached at the beginning; to nurse lepers as he did at the outset."[7] This encapsulates Francis' testimony: the experience of prayer enabled a life marked by reaching out to others.

Julian's Three Windows

Julian of Norwich (c1342–1413) stands within the English mystical tradition. Living as a hermit attached to a parish church, she discovers God through a series of vivid revelations in 1373 which she records and reflects upon. While she glimpses immense suffering in the heart of God as seen on the Cross, her vision of the Divine is supremely optimistic, delighting in a God whose love can conquer all human foibles.

There were thirty anchoresses living in the city of Norwich at this time. Julian dwelt in her cell attached to the parish church of St Julian: "the Lady at St Julian's" became shortened to "the Lady Julian"—we never learn her actual name. Living a life of enclosure, she remained for over twenty years in her small cell, which becomes a place of "Revelation of Divine Love."[8] It

---

6. Merton, *Contemplation*, 263, 264.
7. Bonaventure, "Major Legend", 640. See also Jordan, *Affair of the Heart*, 51, 52.
8. The title of her collected writings. Colledge (ed.), *Julian of Norwich: Showings*.

7 PLANTING A VINE

is a place where she ponders the deepest mysteries of the passion of Christ and dares to call Christ "tender mother." It is a room with three windows, and each opening turns out to be symbolic, representing a different dimension of her vocation.[9] She has a window onto the church, through which she can literally communicate with the Divine, receiving the sacrament of Holy Communion from the altar, which she can glimpse in her view of the sanctuary. This window onto the church represents her longing to be fed and nourished by Christ, food for the spirit.

A second window connects her to her servants and helpers, Alice and Sarah, who attend to her physical needs and bring her the food for the body and requisites for life. This window is a reminder to her of her humanity and indebtedness to others, representing her relationship to others in the Christian family as they help one another out. She cannot survive without this window of practicality. Through this window she receives from others.

But while she makes the cell her abode, she is not "walled up." What is given to her in contemplative prayer is given to her to share. There is a third window—a window onto the world. This curtained window opens onto a little shelter or roofed area where others can come, one by one, to receive counsel and spiritual guidance. Margery Kempe, another local mystic of the time, tells us about this dimension of Julian's ministry and of the spiritual direction she has personally received.[10] To this window come all who seek a word of direction, an insight into their turmoil, an encouragement on the journey: rich and poor, the tradesman and the prostitute, the sailor and the traveler, all are welcome at this window of ministry. Through this window Julian stays in touch with the pains and heartaches of the fourteenth century: devastating upheaval in people's lives, civil unrest, the 100 Years' War. Through this window she learns of the Peasants' Revolt, the Black Death, religious unrest, the burning of Lollards, the great Schism in the western church and the opposing papacies. All this feeds into her prayer of intercession, and all this forms a background to her thinking when she sees in vivid terms the passion of Christ and starts to glimpse the unrelenting and indefatigable love of God. Through this window she listens intently and patiently to what is going on in people's everyday lives. Through this window pours out her compassion and healing as she ministers to the hurting, distraught and confused. This window onto the world represents an opening up in availability to others.

9. I am indebted for this insight from Obbard, *Through Julian's Windows*.
10. Windeatt (trans.), *Margery Kempe*.

Elizabeth Obbard concludes: "Julian can teach us how to balance the conflicting demands on our time, but she can also teach us how to nourish our own inner solitude while befriending others and holding them close in prayer and compassion."[11]

Teresa of Avila: The Prayer of Quiet

In her sketch of the journey of prayer in *The Interior Castle* Teresa of Avila (1515–82) draws us into her fourth room, a place of new discovery which opens us up to "supernatural prayer." Humphreys explains: "Supernatural prayer is where God takes over. It is also called infused contemplation, passive prayer, mystical prayer, or infused prayer. All labels, again, mean the same thing. This type of prayer means that God is communicating with the person."[12] Teresa advises: "if you would progress a long way on this road and ascend to the mansions of your desire, the important thing is not to think much, but to love much; do, then, whatever most arouses you to love."[13] Here Teresa introduces her readers to "the Prayer of Quiet" by means of a powerful picture: while active discursive prayer (using many words and images) can be likened to a basin receiving water from lengthy manmade conduits, pipelines and aqueducts of human effort, the Prayer of Quiet is like a basin placed very close to the spring, at its very source, where the water can flow into it unceasingly and without effort: such is the heart of contemplative prayer. The heart becomes enlarged (Ps 119:32); there is a greater capacity for prayer, a letting-go of former restrictive practices of prayer and a movement from the primacy of ego to the initiative of God.

But, says Teresa, there is no need to rest even here. The *Unitive Way* beckons: we may go deeper into God in Teresa's remaining mansions, which explore different dimensions of contemplative prayer. The fifth mansion is a place of liberation where the soul learns to "fly" in a new freedom.

In a memorable image, Teresa depicts the life of prayer as forming a cocoon around the soul, protecting it from danger. Just as a silkworm works hard to spin its chrysalis, so the Christian, bit by bit, creates by its spiritual disciplines and repeated efforts in prayer, a safe and cosy place in which to reside: "your life is hid with Christ in God" (Col 3:3). However, Teresa insists, this is but a prelude to greater witness to the world, for the soul, like

---

11. Obbard, *Through Julian's Windows*, xvii.
12. Humphreys, *From Ash to Fire*, 80.
13. Peers, *Teresa*, 33.

the caterpillar, must in some sense die and then break free from its self-created imprisonment: "It has wings now: how can it be content to crawl along slowly when it is able to fly?"[14] She urges her readers to break out of their cocoon and realize that they are called to be a beautiful presence of Christ in the world: like a butterfly, the soul is both stunning and vulnerable in this state. In the cocoon, Teresa says "this soul was thinking of nothing but itself" but now it enjoys a wider perspective on the world and discovers a solidarity and shared grief with those who are suffering. The Christian becomes more aware of those in the world who are struggling spiritually. As the soul begins to fly, "here the Lord asks only two things of us: love for His Majesty and love for our neighbor."[15] As the soul breaks free from its self-preoccupation, Teresa advises:

> Ask Our Lord to grant you this perfect love for your neighbor . . . even though this may cause you to forgo your own rights and forget your own good in your concern for theirs . . . try to shoulder some trial in order to relieve your neighbor of it.[16]

The sixth mansion opens the pilgrim-soul to the discovery of a glittering treasury in the inner reaches of the castle. Teresa speaks of the soul's betrothal to God, while in the seventh she uses the daring language of mystical marriage to describe union with God as an abiding awareness and permanent consciousness of unity with the indwelling Christ. But a big question surfaces: is this the goal of the journey within—to rest and luxuriate in intimate communion with Christ? She is emphatic:

> How forgetful this soul, in which the Lord dwells in so particular a way, should be of its own rest . . . For if it is with Him very much, as is right, it should think little about itself. All its concern is taken up with how to please Him more and how or where it will show Him the love it bears Him. *This is the reason for prayer, my daughters, the purpose of the spiritual marriage: the birth always of good works, good works.*[17]

So this experience of union with Christ is not at the cost of total withdrawal from the world. Teresa recalls the story of Martha and Mary to call for an integration between action and contemplation: "This I should like us

---

14. Peers, *Teresa*, 53–55.
15. Peers, *Teresa*, 57, 61.
16. Peers, *Teresa*, 63.
17. Kavanaugh and Rodriguez (trans.), *Interior Castle*, 189, 190.

to attain: we should desire and engage in prayer, not for our enjoyment, but for the sake of acquiring this strength which fits us for service."[18] The Prayer of Quiet leads to costly service of others.

St Benedict: Spirituality of Place

The triple vows of St Benedict's sixth century *Rule* suggest a framework in which to understand and make sense of the sometimes conflicting dynamics of Christian discipleship today. We find ourselves caught between the pull of duty to others and the pull of developing the self, and the first two vows suggest a dialectic within which this tension can be not so much resolved as held creatively.

Stabilitas

Benedict's vow of stability commits the monk to stay with a community for life. It is the first of the vows, because it is about the total surrender of one's life to God, within the particular setting and context of a group of people. In a world where people are rushing around in ever greater degrees of mobility, this commitment invites us to reconsider a rootedness in the here and now, an attentiveness to the needs of a particular community, and firmly rejects the temptation that the grass is greener elsewhere. The vow of *stabilitas* reminds us of the essentially incarnational nature of discipleship, and of the call to be fully present to a particular historical context. But this vow also calls us to consistency, to steadfastness, as Benedict puts it: "to persevere in his stability" (58:9). It calls us to the rock of faithfulness and constancy in a sea of change and tempest of transition. In today's context, this vow is about not giving up, not giving way under the pressures that confront today's Christian, which include marginalization and loss of respect in society. It is about rediscovering God's *hesed* or steadfast love and faithfulness, expressed in Paul's affirmation: "I am confident of this, that the one who began a good work among you will bring it to completion" (Phil 1:6).

Yet the biblical metaphors of "remaining in the vine" (John 15:4) need also to be in conversation with the dynamic language of movement with which Benedict both begins and ends his *Rule*:

---

18. Peers, *Teresa*, 148. See also Bryant, *Journey to the Centre*.

- Run while you have the light of life (Prologue 13)
- Run on the path of God's commandments (Prologue 49)
- Hasten towards your heavenly home (73:8)

Conversatio Morum

And so the vow of *stabilitas* stands in tension with Benedict's second vow of *conversatio morum*, the conversion of life, which calls for constant growth and change. This is an echo of the New Testament call to *metanoia*, turning again to God, and resonates with Paul's resolve: "Straining forward to what lies ahead" (Phil 3:13). Such a commitment can be both liberating and unnerving. It invites us to let go of cherished and familiar ways of working, to be ready for risk-taking, open to experimental and provisional patterns of witness and ministry, which emerge as Christendom dissolves and the church discovers different ways of being in a post-Christian, postmodern world. It calls us to accept the need for lifelong learning and continuous development. It requires of us both a thirst for fresh understanding of God's word and world, and also a vulnerability, a lowering of self-protective barriers, to be open to the God of surprises.

WHAT DOES IT MEAN TO PLANT A VINE IN THE SOUL?

It is to commit ourselves afresh to a central paradox in the spiritual life: abiding and resting in God and ever-open to movements in the soul, shifts in our perception, readiness for costly ministry. The life of discipleship is a constant dance between movement and stillness. We are called to be contemplatives in action—people on the move, while holding in our hearts a center of quietude. The spiritual life is worked out within the interplay between stability and rootedness on the one hand and, on the other, the dynamic call to progress and move on.

PRAYER EXERCISE

Prayerfully reflect on the questions the Vine poses to us:

- What is growing well and blossoming, blooming in my spiritual life?

- Are there buds or potentials that are ready to sprout?
- Do I notice any dead wood that needs to be taken to the fire and burned?
- Can I name attitudes or actions that are evidently unproductive and need to be decisively let go of?
- What parts of my life need pruning or cut back in order to let other parts grow and flourish?
- What sources of inspiration are my spiritual roots exploring?
- What is watering the soul?
- How do I feel when I think of the crushing of the grapes to release their "blood"?

Lord God, whose Son is the true vine and the source of life, ever giving himself that the world may live: may we so receive within ourselves the power of his death and passion that, in his saving cup, we may share his glory and be made perfect in his love; for he is alive and reigns, now and for ever. **Amen.**

BENEATH THE APPLE TREE

8 Planting an Apple Tree

Re-entering Eden

> *As an apple tree among the trees of the wood,*
>   *so is my beloved among young men.*
> *With great delight I sat in his shadow,*
>   *and his fruit was sweet to my taste . . .*
> *Sustain me with raisins,*
> *refresh me with apples;*
> *for I am faint with love . . .*
> *Under the apple tree I awakened you. (Song 2:3,5; 8:5)*

At Borderlands Retreats we have several great apple trees. Indeed, we are surrounded by apple trees on all fronts. Herefordshire is cider country and throughout spring up rich orchards of many different varieties of apple. Outside the west door of Hereford Cathedral in 2011 a mosaic carpet was laid out featuring Christ the Apple Tree, designed by Sandy Elliot, while to mark the Millennium the Leominster community planted an orchard of apple trees in the town, near the Priory—any resident or visitor is invited to pick and enjoy the apples when they come into fruit!

In this chapter we allow ourselves to be guided by the poets. First, as we consider the primordial tree, Traherne will call us to rediscover playfulness in the garden and recover childlike wonder. The eschatological apple tree comes into view as Ephrem in the fourth century dreams of paradise regained. Thirdly, an eighteenth century poem leads us to present joys available right now beneath Christ the apple tree.

1. THE PRIMORDIAL TREE

The apple tree takes us first to Eden. Christian tradition considers that the forbidden fruit in the Garden of Eden taken by Adam and Eve was an apple, popularized in the fifteen-century carol:

> Adam lay ybounden,
>   Bounden in a bond;
> Four thousand winter
>   Thought he not too long.
> And all was for an apple,
>   An apple that he took,
> As clerkës finden written
>   In their book.
>
> Nor had one apple taken been,
>   The apple taken been,
> Then had never Our Lady
>   A-been heaven's queen.
> Blessed be the time
>   That apple taken was.
> Therefore we may singen
>   *Deo gratias!*

The identification of the fruit as apple may be linked to the fact that the Latin word for apple *malum* is almost identical with the word for evil. The carol also expresses the idea of *felix culpa*, as in the Easter Exultet hymn: "O happy fault, O necessary sin of Adam, that gained for us so great a redeemer!"

But let's follow the lead of Herefordshire poet Thomas Traherne (1636–74), who draws us back to the joys and delight of Eden before the Fall. Hereford born and bred, and educated at the city's Cathedral School, he served for the first ten years of his ministry the parish of Credenhill, close to Leominster. Unknown during his lifetime, his *Centuries of Meditations*, reflections on Christian life and his quest for happiness—which he called felicity—were first published in 1908, after having been rediscovered in manuscript only ten years earlier. In 1997 more writings were unearthed in Lambeth Palace Library.[1] Traherne invites us somehow to recover the innocent sense of wonderment that is found in children before they are exposed to the follies and competitiveness of the world. As he reflects on

---

1. Inge, *Happiness and Holiness*.

Eden, he is not interested in sin but in recovering a childlike enchantment with the natural world:

> Certainly Adam in Paradise had not more sweet and curious apprehensions of the world, than I had when I was a child. All appeared new, and strange at first, inexpressibly rare and delightful and beautiful. I was a little stranger, which at my entrance into the world was saluted and surrounded with innumerable joys. My knowledge was Divine. I knew by intuition those things . . . My very ignorance was advantageous. I seemed as one brought into the Estate of Innocence. All things were spotless and pure and glorious: yea, and infinitely mine, and joyful and precious, I knew not that there were any sins, or complaints or laws. I dreamed not of poverties, contentions or vices. All tears and quarrels were hidden from my eyes. Everything was at rest, free and immortal. I knew nothing of sickness or death or rents or exaction, either for tribute or bread. In the absence of these I was entertained like an Angel with the works of God in their splendor and glory, I saw all in the peace of Eden; Heaven and Earth did sing my Creator's praises, and could not make more melody to Adam, than to me: All Time was Eternity, and a perpetual Sabbath. Is it not strange, that an infant should be heir of the whole World, and see those mysteries which the books of the learned never unfold?
>
> The green trees when I saw them first through one of the gates transported and ravished me, their sweetness and unusual beauty made my heart to leap, and almost mad with ecstasy, they were such strange and wonderful things . . .
>
> *But now, with new and open eyes,*
> *I see beneath, as if above the skies,*
> *And as I backward look again*
> *See all His thoughts and mine most clear and plain.*
> *He did approach, He me did woo;*
> *I wonder that my God this thing would do . . .*
>
> Our Savior's meaning, when He said, "You must be born again" and "become a little child to enter into the Kingdom of Heaven" is deeper far than is generally believed. It is not only in a careless reliance upon Divine Providence, that we are to become little children, or in the feebleness and shortness of our anger and simplicity of our passions, but in the peace and purity of all our soul. Which purity also is a deeper thing than is commonly apprehended. For we must disrobe ourselves of all false colors, and unclothe our souls of evil habits; all our thoughts must be infant-like and

clear; the powers of our soul free from the leaven of this world, and disentangled from conceits and customs . . . And therefore it is requisite that we should be as very strangers to the thoughts, customs, and opinions of people in this world, as if we were but little children. So those things would appear to us only which do to children when they are first born. Ambitions, trades, luxuries, inordinate affections, casual and accidental riches invented since the fall, would be gone, and only those things appear, which did to Adam in Paradise, in the same light and in the same colors: God in His works, Glory in the light, Love in our parents, everyone, ourselves, and the face of Heaven: Everyone naturally seeing those things, to the enjoyment of which we are naturally born. [2]

Traherne is inspired by Scriptures that celebrate Eden and paradise regained. The garden of Eden is the primordial image of communion of God, where Adam walks with God in the cool of the day. The exclusion from the garden resonates with our self-imposed exile from God, due to our own personal choices which shatter and fragment the experience of communion with God. There is a nostalgia and longing to return to Eden: it becomes a symbol of a paradise lost that can be regained, as Ezekiel 36: 34, 35 depicts: "The land that was desolate shall be tilled, instead of being the desolation that it was in the sight of all who passed by. And they will say, 'This land that was desolate has become like the garden of Eden.'" Isaiah expresses humanity's longing to return to things as they should be: "For the Lord will comfort Zion; he will comfort all her waste places, and will make her wilderness like Eden, her desert like the garden of the Lord; joy and gladness will be found in her, thanksgiving and the voice of song" (Isa 51:3).

Treharne invites us to let go of worldly attachments that clog the soul and impair a clear vision of the beauty of creation. He longs that we become children once again and allow ourselves to be stunned at God's world. He calls us back to innocence, simplicity and freedom. He tells us that felicity is to be found in such a sense of awe and unfettered, sheer delight at creation:

> That Prospect was the Gate of Heaven; that Day
> The ancient Light of Eden did convey
> Into my Soul: I was an Adam there,
> A little Adam in a Sphere
> Of Joys: O there my ravished Sense
> Was entertained in Paradise;

---

2. Traherne, *Centuries of Meditations*, 67, 69.

> And had a Sight of Innocence
> Which was to me beyond all Price.
> An Antepast of Heaven sure!
> For I on Earth did reign:
> Within, without me, all was pure:
> I must become a Child again.³

A recent poet echoes this capacity for joy in the natural world. In his poem *One Foot in Eden* Edwin Muir celebrates Eden as a place of hope:

> Yet still from Eden springs the root
> As clean as on the starting day.

2. THE ESCHATOLOGICAL TREE

Another poet leads us to contemplate Eden—in a different sense. For Ephrem (306–373), the great Syriac hymnwriter, Eden gives a perfect picture of heaven. The Syriac tradition brims with a bold, imaginative and creative approach to religious language and imagery as it seeks to communicate and put into words the knowledge of the Divine being discovered in prayer. Syriac writers develop an astonishing theological sensitivity to their environment, and discover a spiritual language from the natural world, where physicality points to spirituality.⁴ As a great poet and hymn writer Ephrem pioneered the use of daring language and metaphor. He revealed an astonishing sacramental worldview, in which the whole created order brims with the Divine and teaches us about God's ways.

Ephrem's writings burst out with an awesome variety of symbol and metaphor. For Ephrem the goal of the sanctified life is the recovery of the paradisiacal state, when Adam and Eve were still clothed in the "Robe of Praise." The Robe of glory is a key metaphor. In paradise we will be clothed with a robe of glory in place of the nakedness of Adam and Eve. In the incarnation, God clothes himself in a body. In ongoing divine revelation, God clothes himself in words: "He clothed Himself in language, so that He might clothe us in His mode of life." (*On Faith* 31.2)⁵

---

3. "Innocence" in Wade (ed.), *Poetical Works of Thomas Traherne*, 103.

4. Murray, *Symbols of Church and Kingdom*. He celebrates images in the Syriac tradition borrowed from environment and the natural world, such as vineyard, grape, tree of life, rock.

5. Brock, *Luminous Eye*, 43–44.

The Bible stimulates and encourages both our religious imagination and theological insight:

> Scripture brought me
> to the gate of Paradise,
> and the mind, which is spiritual,
> stood in amazement and wonder as it entered,
> the intellect grew dizzy and weak
> as the senses were no longer able
> to contain its treasures –
> so magnificent they were –
> or to discern its savors
> and find any comparison for its colors,
> or take in its beauties
> so as to describe them in words. (VI:2)[6]
>
> Paradise raised me up as I perceived it,
> it enriched me as I meditated upon it;
> I forgot my poor estate,
> for it had made me drunk with its fragrance.
> I became as though no longer my old self,
> for it renewed me with all its varied nature . . . . (VI:4)
>
> For the colors of Paradise are full of joy,
> its scents most wonderful,
> its beauties most desirable,
> and its delicacies glorious. (IV: 7)
>
> I began to wander
> amid things not described.
> There is a luminous height,
> clear, lofty and fair:
> Scripture named it Eden,
> the summit of all blessings. (V: 5)
>
> There too did I see
> the bowers of the just
> dripping with ointments
> and fragrant with scents,
> garlanded with fruits,
> crowned with blossoms.
> In accord with a person's deeds

---

6. All extracts taken from Brock, *Hymns on Paradise*.

> such was his bower;
> Thus one had few adornments,
> while another was resplendent in its beauty;
> one was dim in its coloring,
> while another dazzled in its glory. (V:6)

> Paradise delighted me
> as much by its peacefulness as by its beauty:
> in it there resides a beauty
> that has no spot;
> in it exists a peacefulness
> that knows no fear.
> How blessed is that person
> accounted to receive it,
> if not by right,
> yet at least by grace . . . (V:12)

Ephrem affirms that the church today, for all its foibles, can be a proleptic image of Paradise:

> God planted the fair Garden,
> He built the pure Church . . . .
> The assembly of saints
> bears resemblance to Paradise:
> in it each day is plucked
> the fruit of Him who gives life to all;
> in it, my friends, is trodden
> the cluster of grapes, to be the Medicine of Life . . . (VI: 7, 8)

The Medicine of Life is a favorite designation in Syriac writing for the Eucharist, an image and anticipation of heavenly joy, which enables us to bear, even here below, a rich harvest of fruits:

> The diligent carry their own fruits
> and now run forward
> to meet Paradise
> as it exults with every sort of fruit.
> They enter the Garden
> with glorious deeds,
> and it sees
> that the fruits of the just
> surpass in their excellence
> the fruits of its own trees . . . (VI:11)

8　PLANTING AN APPLE TREE

We should be inspired and heartened by the example of those who have gone before us:

> The fruits of the righteous
> were more pleasing to the Knower of all
> than the fruits
> and produce of the trees.
> The beauty that exists in nature
> extolled the human mind,
> and Paradise lauded
> the intellect;
> The flowers gave praise to virtuous life,
> and the Garden to free will,
> and the earth to human thought.
> Blessed is He who made Adam so great! (VI:13)
>
> More fitting to be told
> than the glorious account
> of Paradise
> are the exploits of the victorious
> who adorned themselves
> with the very likeness of Paradise;
> in them is depicted
> the beauty of the Garden.
> Let us take leave of the trees
> and tell of the victors . . . (VI: 14)
>
> There in Paradise, manifest and fair,
> to the eye of the mind,
> are the coveted banquets of the just
> who summon us
> to be their brothers and sisters,
> their fellow members and companions.
> Let us not be deprived, my friends,
> of their company;
> let us be their kindred,
> or failing that, their neighbors,
> and if not in their own dwelling,
> at least round about their bowers (VI: 16)

Ephrem speaks directly to those mourning or feeling frail:

> Bear up, O life of mourning,
> so that you may attain to Paradise;
> its dew will wash off your squalor,
> while what it exudes will render you fragrant;
> its support will afford you rest after your toil,
> its crown will give you comfort,
> it will proffer you fruits
> in your hunger,
> fruits that purify those who partake of them ... (V:3)

> Bind up your thoughts, Old Age,
> in Paradise
> whose fragrance makes you young;
> its wafting scent rejuvenates you,
> your stains are swallowed up
> in the beauty which it clothes you (VII:10)

Not only will Paradise in due course renew and rejuvenate us, it will in a sense be a reward for sacrificial living on earth:

> Whoever has washed the feet of the saints
> will himself be cleansed in that dew;
> to the hand that has stretched out
> to give to the poor
> will the fruits of the trees
> themselves stretch out;
> the very footsteps of him
> who visited the sick in their affliction
> do the flowers make haste
> to crown with blooms,
> jostling to see
> which can be first to kiss his steps. (VII: 17)

The trees of Paradise give great welcome, even to those who can't wait to explore the heavenly Eden:

> Should you wish
> to climb up a tree,
> with its lower branches
> it will provide steps before your feet,
> eager to make you recline
> in its bosom above,
> on the couch of its upper branches ... (IX: 3)

> Such is the flowing brook of delights
> that, as one tree takes leave of you,
> the next one beckons to you;
> all of them rejoice
> that you should partake of the fruit of one
> and suck the juice of another,
> wash and cleanse yourself
> in the dew of yet a third;
> anoint yourself with the resin of one
> and breathe another's fragrance,
> listen to the song of still another.
> Blessed is He who gave joy to Adam! (IX:6)

We should not underestimate our capacity for the Divine, even on earth:

> If the beauty of Paradise
> strikes us with astonishment,
> how much more should we be astonished
> at beauty of the mind (VI:15)

Much that Ephrem envisions is an expression of future hope, but he is clear that we have the capacity to enjoy foretastes of Paradise now and, here below, to catch glimpses of the delights that await us:

> The Lord of all
> is the treasure store of all things:
> upon each according to his capacity
> He bestows a glimpse
> of the beauty of His hiddenness,
> of the splendor of His majesty.
>
> As each here on earth
> purifies his eye for Him,
> so does he become more able to behold
> His incomparable glory;
> as each here on earth
> opens his ear to Him,
> so does he come more able to grasp
> His wisdom;
> as each here on earth
> prepares a receptacle for Him,
> so is he enabled to carry
> a small portion of His riches. (IX: 25)

3. THE CONTEMPORARY TREE

In Christian art, Christ has sometimes been depicted as the Second Adam, who even as an infant bears in his hand an apple fruit symbolizing a new creation and a new beginning for humanity—the hope of paradise regained. For example, Paris' Louvre displays the 1489 picture *Virgin and Child Holding an Apple* by an unknown artist from the Southern Netherlands, and *Virgin Mary and Infant Jesus Holding an Apple* by 15th century Boheme. New York's Metropolitan Museum of Art has a *Virgin and Child*, the mother handing an apple to the child, painted by a follower of Hans Memling (Netherlandish, early 16th century), while St Peterburg's Hermitage gallery exhibits *The Virgin and Child Under an Apple Tree* by Lucas Cranach (1472-1553).

But the clearest and best-known image of Christ and this fruit is expressed in the remarkable song dating from the mid-eighteenth century, *Christ the Apple Tree*. Written by Baptist pastor Richard Hutchins, it was first published in America in 1784 in *Divine Hymns, or Spiritual Songs: for the use of Religious Assemblies and Private Christians* compiled by Joshua Smith. It has been set to music by a number of composers, including John Rutter and Elizabeth Poston (1905-1987), whose version is below. *Jesus Christ the Apple Tree* is traditionally sung as a Christmas carol: at Twelfth Night farmers in Britain, from medieval times up until the eighteenth century, would sing and drink to the health of their apple trees—a ritual known as wassailing.

> The tree of life my soul hath seen,
> Laden with fruit and always green;
> The trees of nature fruitless be,
> Compared with Christ the Apple Tree.
>
> His beauty doth all things excel,
> By faith I know but ne'er can tell
> The glory which I now can see,
> In Jesus Christ the Appletree.
>
> For happiness I long have sought,
> And pleasure dearly I have bought;
> I missed of all but now I see
> 'Tis found in Christ the Appletree.
>
> I'm weary with my former toil -

8 PLANTING AN APPLE TREE

Here I will sit and rest awhile,
Under the shadow I will be,
Of Jesus Christ the Appletree.

With great delight I'll make my stay,
There's none shall fright my soul away;
Among the sons of men I see
There's none like Christ the Appletree.

I'll sit and eat this fruit divine,
It cheers my heart like spiritual wine;
And now this fruit is sweet to me,
That grows on Christ the Appletree.

This fruit doth make my soul to thrive,
It keeps my dying faith alive;
Which makes my soul in haste to be
With Jesus Christ the Appletree.

This delightful song celebrates the availability and immediacy of Christ today in this very moment as a contemporary reality. The main emphasis is on the speaker's sensory engagement with this tree—beholding its magnificent stature and lushness, sitting in the coolness of its shade, indulging in its sweet, crisp fruit. It's a love poem—an ode to Christ's beauty, his glory, his truth, and to the pleasure he gives. The poem admires the beauty and fullness of Christ as represented in the tree, always green and weighed down with luscious fruit. The writer reveals why this is such a blessing: he is exhausted from his toil. At last he must escape from the rat-race of life and discover a rejuvenating rest. The shade is comforting, the fruit a necessity for life. It is a place where the poet will return over and over again "with haste."

There is a clear reference to Eden in the first line: "The tree of life my soul hath seen." This is the tree whose fruit that Adam and Eve did not get to enjoy, but is there for us all, right now. We do not have to long with nostalgia for what has been lost, and we do not have to wait until heaven, either, to taste the fruit. We are summoned to take our place beneath the shade of the tree today and taste its rejuvenating fruit in our soul. Jesus is beautiful beyond description and beyond compare. Words are not sufficient to describe him. He is fruitful; abounding in the various fruits of the Spirit, and no one, no thing can compare to his fruitfulness. He is the fount and source of joy and fulfillment. He is providential, providing strength, wisdom, and discernment for those who draw near. The core of Christ's goodness, the stability and constancy of God's goodness, is so perfectly encapsulated in this simple but powerful and moving metaphor. He gives our soul rest, he nourishes our spirit, and he is inherently beautiful beyond words.

The carol speaks of our life-long search for meaning and focus. "For happiness I long have sought, and pleasure dearly I have bought," it reads. Perhaps we never notice the beauty that is right before us. The humble apple tree might pale beside the majesty of an oak; it lacks the grace of an elm and the vibrant colors of a maple. But its small stature offers shade and shelter, and its simple but abundant harvest of fruit provides sustenance deep into the winter.

The last verse offers a message of hope. "This fruit doth make my soul to thrive; it keeps my dying faith alive . . . ." The apple tree becomes a symbol of resurrection, of renewal and fresh beginnings.

WHAT DOES IT MEAN TO PLANT AN APPLE TREE IN THE SOUL?

1. Thomas Traherne invites us to recover playfulness, and childlike wonder at creation.
2. Ephrem tells us that we can not only long for heaven but allow ourselves to be cheered by those bearing fruit on earth. We can learn to see the world sacramentally and resolve and commit to develop habits of perception that enable a new attentiveness and alertness to God's parables in nature.
3. Richard Hutchins' song calls us to delight in the rejuvenating presence of Christ in our lives today.

PRAYER EXERCISE

As you sit beneath the apple tree—in reality or in imagination—ask yourself:

What is my greatest longing and need?

What fruit do I wish to see appear and flourish in my life?

How can I recover playfulness and wonderment?

How far do I allow myself to enjoy sheer delight in Jesus?

End with the prayer from Psalm 17:8
Guard me as the apple of the eye; hide me in the shadow of your wings

"CEDARS OF GOD" RESERVE, MOUNT LEBANON

9 Planting a Cedar
Standing Tall and Ready

The righteous will grow like a cedar in Lebanon.
Planted in the house of the Lord,
They will flourish in the courts of our God. (Ps 92)

TOWERING MAJESTICALLY TOWARDS THE sky, rising in quiet splendor, yet reaching out almost horizontal branches as if to embrace the world, the cedar stands between heaven and earth. One of the most stately of all planted trees with its layers of branches and grey-green foliage, the cedar stands tall amidst the trees of the forest.

It has a double message to us: dignity and availability.

1. DIGNITY

The prophet celebrates its majesty:

> Whom are you like in your greatness?
>   Consider . . . a cedar of Lebanon,
> with fair branches and forest shade,
>   and of great height, its top among the clouds.
> The waters nourished it, the deep made it grow tall,
> making its rivers flow around the place where it was planted,
> sending forth its streams to all the trees of the field.
> So it towered high above all the trees of the field;
> its boughs grew large and its branches long,
>   from abundant water in its shoots.
> All the birds of the air made their nests in its boughs;
> under its branches all the animals of the field
>   gave birth to their young;

and in its shade all great nations lived.
It was beautiful in its greatness, in the length of its branches;
for its roots went down to abundant water.
The cedars in the garden of God could not rival it,
    nor the fir trees equal its boughs;
the plane trees were as nothing compared with its branches;
no tree in the garden of was like it in beauty.
I made it beautiful with its mass of branches,
the envy of all the trees of Eden that were in the garden of God. (Ezek 31:3–9)

The cedar is celebrated for its height, beauty, sturdiness and strength. It becomes a symbol of the noble soul. The psalmist prays:

> The righteous will flourish like the palm tree,
> They will grow like a cedar in Lebanon.
> Planted in the house of the Lord,
> They will flourish in the courts of our God.
> They will still yield fruit in old age;
> They shall be full of sap and very green (Ps 92:12–15, NASB1995)

It comes to express the destiny and vocation of God's people. Ezekiel uses this image to speak of the renewal and restoration of his exiled people:

> Thus says the Lord GOD:
> I myself will take a sprig
> from the lofty top of a cedar;
>   I will set it out.
> I will break off a tender one
>   from the topmost of its young twigs;
> I myself will plant it
>   on a high and lofty mountain.
> On the mountain height of Israel
>   I will plant it,
> in order that it may produce boughs and bear fruit,
>   and become a noble cedar.
> Under it every kind of bird will live;
>   in the shade of its branches will nest
>   winged creatures of every kind.
> All the trees of the field shall know
>   that I am the LORD. (Ezek 17: 22–24)

Ezekiel celebrates the potential and the beauty of the noble cedar, given by God. Our dignity, our identity does not come from what other people say about us. It comes from what God says about us—and he declares to us, in the waters of baptism: " You are my beloved son, my cherished daughter."

Spiritual writers urge us to rejoice in our true identity and possibilities.

In 1577 Teresa of Avila cries out to her sisters: "O souls redeemed by the blood of Jesus Christ! Learn to understand yourselves!" And she goes on: "In speaking of the soul we must always think of it as spacious, ample and lofty; and this can be done without the least exaggeration, for the soul's capacity is much greater than we can realize!" [1]

Recent writers, too, affirm: "The most courageous thing we will ever do is to bear humbly the mystery of our own reality."[2] How do you find yourself responding to the prayer: "God, help me to believe the truth about myself, no matter how beautiful it may be"?[3] Respond with both your head and your heart!

2. AVAILABILITY

In the passages from Ezekiel, wanting to express hope and confidence in a new future that is beckoning, the prophet sees the image of the cedar as a suitable metaphor to convey new beginnings, resilience and openness to others. Outstretched branches welcome birds of every kind, making the cedar as a symbol of a preparedness to embrace others. The cedar's extending branches, like outreached arms, seem to speak of availability, of openness. It is a characteristic of the cedar that it can be employed for sacred purposes.

Building a Temple

The use of cedar in building the temple represents the consecrated use of wood—not exploitative as in Solomon's House of the Forest of Cedars, a military store room, as we noted in chapter 1. But the temple was different. Cedar brought from Lebanon provides beams and paneling: this was cedar used reverentially, dedicated, given wholly over to God, that it might furnish a fitting dwelling place of the symbol of divine presence, the Ark of the Covenant. The idea of a temple came first to David: "the king said

---

1. Peers, *Teresa*, 6, 8.
2. Rohr, *Everything Belongs*, 97.
3. Wiederkehr, *Seasons*.

to the prophet Nathan, 'See now, I am living in a house of cedar, but the ark of God stays in a tent'" (2 Sam 7:2).

Solomon oversaw the construction of the temple in about 1000 BC.

> Solomon built the house, and finished it. He lined the walls of the house on the inside with boards of cedar ... He built twenty cubits of the rear of the house with boards of cedar from the floor to the rafters, and he built this within as an inner sanctuary, as the most holy place. The house, that is, the nave in front of the inner sanctuary, was forty cubits long. The cedar within the house had carvings of gourds and open flowers; all was cedar, no stone was seen. The inner sanctuary he prepared in the innermost part of the house, to set there the ark of the covenant of the LORD. The interior of the inner sanctuary was twenty cubits long, twenty cubits wide, and twenty cubits high; he overlaid it with pure gold. He also overlaid the altar with cedar. (I Kgs 6:14–20)

The day of dedication was memorable: "When Solomon had ended his prayer, fire came down from heaven and consumed the burnt-offering and the sacrifices; and the glory of the LORD filled the temple" (2 Chron 7).

Here, it might be said, the cedar reaches its fullest potential. It is cut down, sacrificed as a tree but for the most holy purpose imaginable. This points forward to Paul's understanding, that the human body—like so much cedar wood—can become a living sanctuary and dwelling place for the Divine:

> Do you not know that you are God's temple and that God's Spirit dwells in you? (1 Cor 3:17)

> Do you not know that your body is a temple of the Holy Spirit within you, which you have from God, and that you are not your own? (1 Cor 6:19)

> In Christ Jesus the whole structure is joined together and grows into a holy temple in the Lord; in whom you also are built together spiritually into a dwelling-place for God. (Eph 2:19–22)

Cleansing and Purifying

Jesus himself referred to the cleansing potential of cedar in the episode of the healing of the leper:

> Jesus stretched out his hand, touched him . . . Immediately the leprosy left him. And he ordered him to tell no one. "Go", he said, "and show yourself to the priest, and, as Moses commanded, make an offering for your cleansing, for a testimony to them." (Luke 5:13,14)

Jesus was referring to the passages in Leviticus:

> The LORD spoke to Moses, "These are the instructions for making a person clean after a skin disease. He must be taken to the priest. The priest will go outside the camp and examine him. If the person is healed, the priest will order someone to get two living, clean birds, some cedar wood, red yarn, and a hyssop sprig to use for the cleansing . . .
>
> The priest must take two birds, cedar wood, red yarn, and a hyssop sprig and use them to make the house clean. (Lev 14:1–4, 50)

The Book of Numbers adds: "The priest shall take cedar wood, hyssop, and crimson material, and throw them into the fire of sacrifice" (19:6). These ceremonies involving the use of cedar were designed to being public affirmation and reassurance to the person, restored both in themselves and in their place in the community.

Healing

Through the prophet Hosea, God declares:

> I will be like the dew to Israel;
> My people will blossom like the lily,
> And take root like the cedars of Lebanon.
> Their shoots will sprout,
> And their beauty will be like the olive tree,
> their fragrance like the cedars of Lebanon. (Hos 14:5–6)

The prophet refers to the aromatic character of cedar wood. So too does the Bridegroom in the Song of Songs:

> How sweet is your love, my sister, my bride!

> how much better is your love than wine,
> and the fragrance of your oils than any spice!
> Your lips distil nectar, my bride;
> honey and milk are under your tongue;
> the scent of your garments is like the scent of Lebanon. (Song 4:10,11)

Cedar oil is produced from the foliage, wood, roots, and bark of cedar. Cedarwood oil has many uses in medicine and perfumery. It has been in use since the time of the ancients, first used by Sumerians and Egyptians.

Used medicinally, cedarwood essential oil protects the body against harmful bacteria, facilitates wound-healing, soothes the discomforts of muscle aches, joint pain or stiffness, calms coughs as well as spasms, supports the health of the organs, and stimulates circulation. Cedar bark has anti-inflammatory properties, and was frequently applied as a dressing for wounds, as a tourniquet, or to ward off evil. So cedar has been used as a medicine for both the body and the spirit. Some Native American Northwest tribes used red cedar for such things as stomach pain relief, tuberculosis, fever reducer, arthritis, and many other medicinal purposes. It is widely used today in aromatherapy.

Making Music

> Praise him with trumpet sound;
> praise him with lute and harp!
> Praise him with tambourine and dance;
> praise him with strings and pipe! (Ps 150:3,4)

In the worship of ancient Israel these musical instruments had a vital part in "making a joyful noise unto the Lord" and cedar wood played a key role in enabling this to happen.

The lute is a kind of guitar, often made from cedars which help produce a classic warm, rich and resonant tone. The soundboard of harps are also made from a coniferous softwood like cedar which gives an excellent tonal response. Today the highly resonant cedar is the most popular top tonewood for classical guitars—and is an excellent choice for steel-string acoustics as well—not known in ancient Israel!

9 PLANTING A CEDAR

MEDITATING WITH THE CEDAR IN NATURE

Cedar trees were a vital part of the stunning landscape that surrounded the hermitage of mystic and theologian Thomas Merton (1915–1968) at the Trappist monastery of Gethsemani in Kentucky.

> Out here in the woods, I can think of nothing except God ... I am as aware of God as of the sun and the clouds and the blue sky and the thin cedar trees ... Engulfed in the simple and lucid actuality which is the afternoon: I mean God's afternoon, this sacramental moment of time when the shadows will get longer and longer, and one small bird sings quietly in the cedars, and one car goes by in the remote distance and the oak leaves move in the wind. High up in the late summer sky I watch the silent flight of a vulture, and the day goes by in prayer. This solitude confirms my call to solitude. The more I am in it, the more I love it.[4]

Walking in the vicinity of his monastery, one day in December 1949, he came to a new and vivid awareness:

> These clouds low on the horizon, the outcrops of hard yellow rock on the road, the open gate, the perspective of fence posts leading up the rise to the sky, and the big cedars tumbled and tousled by the wind. Standing on rock. Present. The reality of the present and of solitude divorced from past and future ... My love for everybody is equal, neutral and clean. No exclusiveness. Simple and free as the sky, because I love everybody and am possessed by nobody, not held, not bound. In order to be remembered or even wanted I have to be a person that nobody knows ... For my part my name is that sky, those fence posts, and those cedar trees ...[5]

MEDITATING WITH THE CEDAR IN SCRIPTURE

The great Baptist preacher Charles Spurgeon (1834–92) delighted in the cedar, seeing it as an image of the Christian:

> "The cedars of Lebanon which he hath planted." (Ps 104:16)

> Lebanon's cedars are emblematic of the Christian, in that we owe our planting entirely to the Lord. This is quite true of every child of

---

4. Merton, *Search for Solitude*.
5. Merton, *Sign of Jonas*.

God. We are not man-planted, nor self-planted, but God-planted. The mysterious hand of the divine Spirit dropped the living seed into a heart which he had himself prepared for its reception. Every true heir of heaven owns the great Husbandman as his planter.

Moreover, the cedars of Lebanon are not dependent upon man for their watering; they stand on the lofty rock, unmoistened by human irrigation; and yet our heavenly Father supplieth them. Thus it is with Christians who have learned to live by faith. We are independent of man, even in temporal things; for our continued maintenance we look to the Lord our God, and to him alone. The dew of heaven is our portion, and the God of heaven is our fountain.

Again, the cedars of Lebanon are not protected by any mortal power. They owe nothing to man for their preservation from stormy wind and tempest. They are God's trees, kept and preserved by him, and by him alone. It is precisely the same with Christians. We are not hot-house plants, sheltered from temptation; we stand in the most exposed position; we have no shelter, no protection, except this, that the broad wings of the eternal God always cover the cedars which he himself has planted. Like cedars, believers are full of sap, having vitality enough to be ever green, even amid winter's snows.

Lastly, the flourishing and majestic condition of the cedar is to the praise of God only. The Lord, even the Lord alone hath been everything unto the cedars, and, therefore David very sweetly puts it in one of the psalms, "Praise ye the Lord, fruitful trees and all cedars." In believers there is nothing that can magnify man; we are planted, nourished, and protected by the Lord's own hand, and to him let all the glory be ascribed.[6]

WHAT DOES IT MEAN TO PLANT A CEDAR IN THE SOUL?

The once-mighty Cedar of Lebanon forests of antiquity have been almost entirely eradicated, but a program of re-planting is in hand in Lebanon. What might our spiritual replanting symbolize?

---

6. Spurgeon, "Trees in God's Court."

Reclaim your Dignity

Our sense of self-worth can easily get eroded by different factors in our culture. The cedar beckons us to pause and rejoice in our God-given attributes, our capacity for the Divine—we are made in the image and likeness of God himself! Like the cedar, you are magnificent!

Re-affirm your Readiness

As the cedar stands ready to be used for sacred purposes—but not for selfish exploitation –so it invites us to take time to reflect afresh on our vocation to be instruments in God's hands.

PRAYER EXERCISE

This exercise is in two parts.

First stand (or sit) upright. Reach yourself up to your full height. Hold your head high—not in pride but in confidence in God. Celebrate your dignity as a child of God. Whisper to yourself the baptismal affirmation: "I am God's beloved son . . . I am God's cherished daughter—with me he is well-pleased." Remind yourself that in God's sight you are like a noble cedar, beautiful, and, in him, strong. Smile that God delights in you. Repeat to yourself the words of John: "See what love the Father has lavished on us in letting us be called God's children! For that is what we are!" (1 John 3:1, CJB). Let any doubts about your dignity as God's beloved evaporate. Hold on to this truth, and let it sink from head to heart.

Second, now open your arms wide, like the horizontal branches of the cedar. Surrender yourself to God, as cedar wood was yielded to God's purposes, in building a temple, in healing, in making music. To what is God now calling you? Affirm your availability to him, saying "take me and use me to your glory."

PLANTING AN OLIVE

10   Planting an Olive

Rediscovering the Cross

He himself bore our sins in his body on the tree, that we might die to sin and live to righteousness. By his wounds you have been healed. (1 Peter 2:24, RSV)

THE TREE OF THE CROSS

ITS SILVERY GREEN BRANCHES sparkle in the sun and its knotted trunk with smooth, ash-colored bark, rises majestically from the earth. The olive tree, in all its mystery and majesty, draws us to contemplate the Cross.

The writers of the New Testament often say that Jesus was crucified on a tree:

> When they had carried out everything that was written about him, they took him down from the tree and laid him in a tomb. (Acts 13:29)

> Christ redeemed us from the curse of the law by becoming a curse for us—for it is written, "Cursed is everyone who hangs on a tree." (Gal 3:13)

It is not known from what tree the wood of the cross was hewn. In a broad valley west of Jerusalem the Monastery of the Holy Cross, dating from the fifth century, preserves one tradition. Medieval frescoes on its walls tell the story of Abraham receiving from his angelic visitors (Gen 18:1–15) three staffs, of cedar, cypress and pine. After Sodom is destroyed, Abraham gives the staffs to his nephew Lot, who plants them and waters them from the Jordan River. The three woods grow into a single tree, and centuries later the tree is cut down and a beam prepared for the cross.

Modern archaeology in fact tells us that some people were crucified on olive wood crosses. In 1968 in an ossuary discovered at Giv'at ha-Mivtar near Jerusalem contained the heel bone of one Yehohanan, dated to the first century. It had been pierced through in crucifixion with an iron nail, still in situ, and furthermore the nail had olive wood particles around it, suggesting he was hung either on a crude olive cross or tree. At Borderlands Retreats we have planted a small olive grove—yes, even in England!—and there retreatants sit to ponder the wood of the cross.

Three texts will inspire our prayers.

First we will look at the sixth century hymns by Fortunatus, so important in the western church.

Then we will enjoy the Old English poem *The Dream of the Rood*, parts of which date from the eighth century.

Finally, we look at the Franciscan Bonaventure's thirteenth century meditation *The Tree of Life*.

Above all, we shall ask: what does it mean to plant an olive tree in your soul?

1. CRUX FIDELIS

Italian poet Venantius Fortunatus (530–609) composed several lyrics on the holy cross, and we shall reflect on two of them. The first, *Crux Fidelis*, brims with wonder at the unique vocation of this tree: "None in foliage, none in blossom, none in fruit thy peers may be."

> Faithful Cross!
> above all other,
> one and only noble Tree!
> None in foliage, none in blossom,
> none in fruit thy peers may be;
> sweetest wood and sweetest iron!
> Sweetest Weight is hung on thee!
>
> Lofty tree, bend down thy branches,
> to embrace thy sacred load;
> oh, relax the native tension

of that all too rigid wood;
gently, gently bear the members
of thy dying King and God.

Tree, which solely was found worthy
the world's Victim to sustain.
Harbor from the raging tempest!
Ark, that saved the world again!
Tree, with sacred blood anointed
of the Lamb for sinners slain.

Blessing, honor, everlasting,
to the immortal Deity;
to the Father, Son, and Spirit,
equal praises ever be;
glory through the earth and heaven
to Trinity in Unity. Amen. (trans. J.M. Neale)

Vexilla regis prodeunt, was composed by Fortunatus to commemorate the sending of a relic of the True Cross from Emperor Justin II to St. Radegund. The hymn was sung as the relic was carried from Tours to Radegund's monastery at Poitiers in the year AD 569. But its words came to be widely appreciated, the hymn finding its way, for example, into the Methodist Hymn Book and the Hymn Book of the Evangelical Association.

The royal banners forward go;
The cross shows forth redemption's flow,
Where He, by whom our flesh was made,
Our ransom in His flesh has paid:

Where deep for us the spear was dyed,
Life's torrent rushing from His side,
To wash us in the precious flood
Where flowed the water and the blood.

Fulfilled is all that David told
In sure prophetic song of old.
That God the nations' king should be
And reign in triumph from the tree

On whose hard arms, so widely flung,
The weight of this world's ransom hung,
The price of humankind to pay

And spoil the spoiler of his prey.

O Tree of beauty, tree most fair,
Ordained those holy limbs to bear:
Gone is thy shame, each crimsoned bough
Proclaims the King of Glory now.

To Thee, eternal Three in One,
Let homage meet by all be done;
As by the cross Thou dost restore,
So guide and keep us evermore. (trans. J. M. Neale)

These hymns direct the worshipper to the wood of the Cross in a sense of wonderment and amazement: "one and only noble Tree", "Tree of beauty, tree most fair." They invite us to be awed by the Cross—not to try to work out a theory of the atonement, nor to count our sins, but to rediscover the prayer of adoration.

What hymn to the Tree of the Cross can you compose?

2. THE DREAM OF THE ROOD

Preserved in the tenth century Vercelli Book, parts of this poem have been found on the eighth century Ruthwell Cross (now in Scotland but once part of the Anglo-Saxon kingdom of Northumbria). One of the oldest works of Old English literature, it may represent an attempt by Anglo-Saxon monks to christianize pagan devotion to trees.

The poem is unique. The first part describes a dream about the tree that became the cross—it appears first as bejeweled with gems and then becomes red with the flow of blood upon it. In the second part—we have an extract below—the Tree speaks and shares its experience of becoming wounded, even as it receives the body of Jesus. Neither Jesus nor the Cross is given the role of the helpless victim in the poem, but instead both stand firm. Jesus is depicted as the strong conqueror, a young warrior, indeed a hero who achieves a strange victory. In the third part the poet celebrates the joy and hope of this victory and looks forward to its fulfilment in heaven.

> Lo! I will tell the dearest of dreams
> That I dreamed in the midnight when mortal men
> Were sunk in slumber . . .
> Then, as I lay there, long I gazed

In rue and sadness on my Savior's Tree,
Till I heard in dream how the Cross addressed me,
Of all woods worthiest, speaking these words:

"Long years ago (well yet I remember)
They hewed me down on the edge of the holt,
Severed my trunk; strong foemen took me,
For a spectacle wrought me, a gallows for rogues.
High on their shoulders they bore me to hilltop,
Fastened me firmly, an army of foes!

"Then I saw the King of all mankind
In brave mood hasting to mount upon me.
Refuse I dared not, nor bow nor break,
Though I felt earth's confines shudder in fear;
All foes I might fell, yet still I stood fast.

"Then the young Warrior, God, the All-Wielder,
Put off His raiment, steadfast and strong;
With lordly mood in the sight of many
He mounted the Cross to redeem mankind.
When the Hero clasped me I trembled in terror,
But I dared not bow me nor bend to earth;
I must needs stand fast. Upraised as the Rood
I held the High King, the Lord of heaven.
I dared not bow! With black nails driven
Those sinners pierced me; the prints are clear,
The open wounds. I dared injure none.
They mocked us both. I was wet with blood
From the Hero's side when He sent forth His spirit.

"Many a bale I bore on that hillside
Seeing the Lord in agony outstretched.
Black darkness covered with clouds God's body,
That radiant splendor. Shadow went forth
Wan under heaven; all creation wept
Bewailing the King's death. Christ was on the Cross.

"Then many came quickly, faring from far,
Hurrying to the Prince. I beheld it all.
Sorely smitten with sorrow in meekness I bowed
To the hands of men. From His heavy and bitter pain
They lifted Almighty God. Those warriors left me

Standing bespattered with blood; I was wounded with spears.
Limb-weary they laid Him down; they stood at His head,
Looked on the Lord of heaven as He lay there at rest
From His bitter ordeal all forspent. In sight of His slayers
They made Him a sepulcher carved from the shining stone;
Therein laid the Lord of triumph. At evening tide
Sadly they sang their dirges and wearily turned away
From their lordly Prince; there He lay all still and alone.

"There at our station a long time we stood
Sorrowfully weeping after the wailing of men
Had died away. The corpse grew cold,
The fair life-dwelling. Down to earth
Men hacked and felled us, a grievous fate!
They dug a pit and buried us deep.
But there God's friends and followers found me
And graced me with treasure of silver and gold.

"Now may you learn, O man beloved,
The bitter sorrows that I have borne,
The work of caitiffs. But the time is come
That men upon earth and through all creation
Show me honor and bow to this sign.
On me a while God's Son once suffered;
Now I tower under heaven in glory attired
With healing for all that hold me in awe.
Of old I was once the most woeful of tortures,
Most hateful to all men, till I opened for them
The true Way of life. Lo! the Lord of glory,
The Warden of heaven, above all wood
Has glorified me as Almighty God
Has honored His Mother, even Mary herself,
Over all womankind in the eyes of men . . . [1]

How do you find yourself responding to the Dream?

Do you feel any empathy with the tree used for crucifixion, as in the *Dream of the Rood*? Can you identify with it in any way?

If you were the tree receiving the offering of Jesus, what would you want to say—To him? To the world? To yourself?

1. Kennedy, *Poems*, 62–65.

3. THE TREE OF LIFE

Franciscan Bonaventure (1221–1274) calls us to plant a cross-shaped tree in the center of our being:

> *With Christ I am nailed to the* cross (Gal 2:20)
> As a true worshipper of God and disciple of Christ,
> who desires to conform perfectly
> to the Savior of all
> crucified for you,
> above all, strive
> with an earnest endeavor of soul
> to carry about continuously,
> both in your soul and in your body,
> the cross of Christ
> until you can truly feel in yourself
> what the Apostle said above. (119) [2]

> Imagination aids understanding . . . Picture in your mind a tree whose roots are watered by an ever-flowing fountain that becomes a great and living river with four channels to water the garden of the entire Church.

> Imagine that from the trunk of this tree there are growing twelve branches that are adorned with leaves, flowers and fruit.

> Imagine that the leaves are a most effective medicine to prevent and cure every kind of sickness, because the wood of the cross is "the power of God for salvation to everyone who believes" (Rom 1:16). Let the flowers be beautiful with the radiance of every color and perfumed with the sweetness of every fragrance, awakening and attracting the anxious hearts of people of desire.

> Imagine that there are twelve fruits, "having every delight and the sweetness of every taste" (Wis 16:20). This fruit is offered to God's servants to be tasted so that when they eat it, they may always be satisfied, yet never grow weary of its taste. This is the fruit that took its origin from the Virgin's womb and reached its savory maturity on the tree of the cross under the midday heat of the eternal Sun, that is, the love of Christ. In the garden of the heavenly paradise—God's table—this fruit is served to those who desire it.

---

2. Page numbers are extracts from Bonaventure, *Soul's Journey into God*.

The lower branches witness to the origin of the savior and his humble life.

The middle branches witness to his passion.
The upper branches witness to his glorification.

Within the middle section four fruits are offered to us:
Christ's confidence in time of trial
Christ's patience when mistreated
Christ's constancy under torture
Christ's victory in the conflict of death

Christ's Confidence in Time of Trial

>Ruler, Lord Jesus,
>From where comes to your soul [in Gethsemane]
>such vehement anxiety and such anxious supplication?
>To shape us in faith
>by believing that you have truly shared our mortal nature,
>to lift us up in hope
>when we must endure similar hardships,
>to give us greater incentives to love you –
>for these reasons you exhibited
>the natural weakness of the flesh
>by evident signs which teach us
>that you have truly borne our sorrows (Isa 53:4)
>and that it was not without experiencing pain
>that you tasted the bitterness
>of your passion. (142)

Christ's Patience when Mistreated

>O truthful and kind Jesus,
>what soul who is devoted to you,
>when it sees and hears this,
>can restrain itself from tears
>and hide the sorrow
>of its inner compassion? (145)

Christ's Constancy under Torture

> O words full of sweetness and forgiveness:
> "Father, forgive them!"
> O words full of love and grace:
> "Today you will be with me in Paradise!"
> Breathe in peace now, O soul,
> in hope of pardon ... (150)

Christ's Victory in the Conflict of Death

> Rise, therefore beloved of Christ,
> be like the dove
> that makes its nest in the mouth of a cleft (Jer 48:28).
> There,
> like a sparrow that finds a home (Ps 83:4),
> do not cease to keep watch;
> there,
> like a turtledove,
> hide the offsprings of your chaste love;
> there
> apply your mouth
> to draw water from the Savior's fountains (Isa 12:3)
> for this is the river
> arising from the midst of paradise ...
> and flowing into devout hearts,
> waters and makes fertile
> the whole earth. (155)

Which of the four fruits of the passion do you need most? Why?

WHAT DOES IT MEAN TO PLANT AN OLIVE IN THE SOUL?

It is to move from one state to another:

- from pain and to hope
- from blood to beauty
- from failure to faithfulness

- from desolation to dedication
- from death to resurrection and newness of life

We celebrate the possibilities as we take on our lips the words of the eucharistic prayer:

> And now we give you thanks because, for our salvation, he was obedient even to death on the cross. The tree of shame was made the tree of glory; and where life was lost, there life has been restored. (Common Worship)

PRAYER EXERCISE

Use the "cross-prayers" devised by Francis of Assisi.

1. Open your arms wide—extend them as far as you can. This is first to embody a solidarity with the cross. Think of Jesus opening wide his arms on the cross to embrace all who suffer, all who are in any form of distress. Think of Christ's all-encompassing love and acceptance. Recall too Jesus' image of the tree that opens its branches wide to welcome birds, as a symbol of outreaching and enfolding love (Luke 13:18,19).

2. Second, think of the Risen Christ and the way he longs to enfold the whole world, the little ones and marginalized ones of the earth.

3. Third, offer this prayer as an act of intercession. It is a prayer that hurts—in the sense that your arms will grow weary and ache. Moses prayed like this and had to have others hold his arms up (Exod 17:11,12). As you feel the ache, let it connect you to those who are in pain, those who are hurting: the sick, the dispossessed, those whose human rights are trampled on.

4. Finally, use this prayer-action as an act of self-offering. Offer yourself afresh to God for the part he has in store for you in his mission of reconciliation in the world. Close your prayer time with this modern hymn:

> Cross of Christ, immortal tree
> On which our Savior died,
> The world is sheltered by your arms

That bore the Crucified.

From bitter death and barren wood
the tree of life is made;
Its branches bear unfailing fruit
And leaves that never fade.

O faithful Cross, you stand unmoved
While ages run their course;
Foundation of the universe,
Creation's binding force.

Give glory to the risen Christ
And to his Cross give praise,
The sign of God's unfailing love,
The hope of all our days. (Stanbrook Abbey,1974)

YEW TREE AT LEOMINSTER PRIORY

11 Planting a Yew Tree

Regenerating

> *Happy are those who find wisdom,*
> *and those who get understanding...*
> *She is a tree of life to those who lay hold of her;*
> *those who hold her fast are called happy. (Prov 3:13, 18)*

HEREFORDSHIRE YEWS

THE YEW IS ONE OF THE most ancient trees in the world—some sources say that it is the oldest-living one. This makes it an incredibly important tree when considering creation and time itself. Appropriately, the yew tree is symbolic of immortality and everlasting life, rebirth, changes and regeneration after difficult times.

Britain has the largest collection of ancient yew trees on earth and here at Borderlands Retreats yews seem to spring up everywhere! We have three main trees, but sprigs emerge all over the place, a veritable sign of new growth and fresh beginnings. Herefordshire is famous, not only for its apple trees (and cider) but also for its ancient yews, all in churchyards. Before the advent of Christianity, yew trees were a focus for pagan (the word means "country-loving") Celtic worship. The Romans called one continental tribe the Eburones, "the people of the yew" after the places where they worshipped. The yew tree was both a focus for devotion, because of its association with longevity and fertility, and a meeting place for the community. As Gregory the Great had advised Augustine in his mission to England in 597 to plant churches in relation to existing devotional sites, so the first churches in England were often established in sacred groves of yews, where people had already sought the Divine for centuries. Pagan holy places were not to be destroyed, but to be converted to Christian use, as pagan feasts were to be superseded by Christian ones. A yew tree is said to have sheltered Augustine when he arrived in England to bring the faith there.

Living longer than any other British native tree species, yew trees can reach 2000 years of age. Indeed the Fortingall yew set within a churchyard in Perthshire in Scotland may be three thousand years old. The name of the holy island of Iona is derived from Ioua, the Pict's word for yew.

Three ancient yews in Herefordshire predate the coming of the Christian faith.

People in Peterchurch had always believed that the yew close to St Peter's Church was 750 years-old, but recent research revealed that in fact it is one of the oldest in the country approaching 3000 years old. The yew tree is located on the north side of the church in the spot where the neolithic people used to mark their burial grounds.[1]

Near Ross on Wye, the massive Linton Yew is thought to be one of the oldest yew trees in Britain. The largest of two old yews in St Mary's churchyard, Herefordshire, this female tree has a very wide girth, and a huge, hollowed out trunk. Like all ancient yews, it is very difficult to accurately date the Linton Yew, especially as there are no rings to count in its hollowed trunk but it is estimated at about 1,500 years old. The Linton Yew was severely damaged by fire in 1998 but the English yew (*Taxus baccata*) has an amazing capacity to regenerate, and it has recovered well, producing new shoots all over its trunk and branches. A wonderful example of yew regeneration, the Linton Yew has a five foot internal stem, while secondary wood flows over the old white wood on both sides of the cavity that faces the path to the church porch.

Much Marcle's yew is believed to be 1500 years old. The 13th century church is just the latest companion of the ancient yew found in the churchyard: it has seen a medieval castle and Norman church, fragments of which are still visible. The yew has hollowed and a bench placed inside it.[2]

---

1. Goddard, "Churchyard Yew Tree."
2. Wills, *History of Trees.*

11   PLANTING A YEW TREE

Sadly over five hundred ancient yew trees have been destroyed in the UK since the Second World War, often chopped down into pieces and stacked waiting to be turned into firewood. The time has come for urgent replanting.

The yew is not mentioned in the bible as such, but its very close relatives, part of the same conifer family, juniper (sometimes known as the mountain yew) and cypress are mentioned. God's enduring faithfulness is likened to an evergreen: "I am living and strong! I look after you and care for you. I am like an evergreen tree, yielding my fruit to you throughout the year" (Hos 14:8, *Living Bible*).

> I will heal their faithlessness;
>   I will love them freely,
>   for my anger has turned from them.
> I will be as the dew to Israel;
>   he shall blossom as the lily,
>   he shall strike root as the poplar;
> his shoots shall spread out . . .
>
> O Ephraim, what have I to do with idols?
>   It is I who answer and look after you.
> I am like an evergreen cypress,
>   from me comes your fruit.
> Whoever is wise, let him understand these things;
>   whoever is discerning, let him know them;
> for the ways of the Lord are right,
>   and the upright walk in them,
>   but transgressors stumble in them. (Hos 14:4–9)

The renewal of God's people is symbolized in the cypress flourishing once again:

> Instead of the thorn shall come up the cypress;
>   instead of the brier shall come up the myrtle; (Isa 55:12)

WHAT SUSTAINS RENEWAL?

WHAT DOES IT MEAN TO PLANT A YEW TREE OR CYPRESS IN THE SOUL?

As this retreat draws to its end, the yew invites us to take stock of our spiritual life—making some kind of inner audit—so we can celebrate God's providence and work in us thus far and move confidently into the future. First we reflect on seven key elements and then respond to seven invitations from the yew . . .

7 VITAL FEATURES

1. Roots

While the root systems of different yews vary, many have deep, widespread roots. The root system is ever-expanding both in depth and in extent as the tree searches out fresh sources of nourishment. It becomes able to draw up a variety of life-sustaining nutrients. Of course as roots seek to progress they will face unseen impediments and barriers like rocks blocking their subterranean pathways.

This encourages us to spread our spiritual roots both wide and deep. We are beckoned to become curious, explorative, deep-rooted "that Christ may dwell in your hearts through faith; that you, being rooted and grounded in love, may be able . . . to know the love of Christ" (Ephesians 3:17–19 NKJV).

How deep are your spiritual roots? How wide are they? Is there anything holding you back from being ever more curious and inquisitive in your spiritual journey? Are you facing any resistances or barriers getting in the way of your progress? What strategy might you develop to overcome these?

2. Seeds

Yew seeds are disseminated both by the wind and by the agency of birds. This means that fresh sprigs can emerge and flourish all over the place. We must allow the breeze of the Spirit to blow through our lives, and become ever more be generous in sharing what we have. We are challenged not to hoard our spiritual reserves in a self-protective way but rather be outgoing in our readiness to allow our seeds to be scattered.

What seeds do you have to share, as a result of this retreat?

3. Branches

Drooping branches of old yew trees can root and form new trunks where they touch the ground. Yew trees were seen as being immortal, able to grow their branches down into the earth to form new trees. The yew's branches can actually grow into the ground so, when the trunk dies, the tree is able to continue living. The yew becomes a symbol of continuity and connection between the past and the present, particularly in terms of having a relationship with one's spiritual heritage.

What continuities can you trace and celebrate from the lives of older Christians who have inspired you?

What fresh saplings are sprouting up from your life of discipleship?

4. Leaves

The leaves of the yew glisten and shimmer in the sunlight as they drink in the energy of the sun. On mature yews the little leaves grow everywhere—up the trunk, along the length off branches and not just at the end of them. We need to catch the light of Christ in a daring variety of ways and rule nothing out. Maximize your opportunities to catch the light of Christ, to receive the light. Make the most of every moment.

How can you become more receptive each day to God's light?

Leaves can be gentle but strong, vulnerable and easily torn, fragile but resilient, structural yet responsive. A leaf might seem insignificant in itself, but working with others, is the lungs of the earth.

Go outside and pray with the foliage you see and touch. What does it say to the paradoxes of your life?

5. Fruits

Anti-cancer compounds are harvested from the foliage of the common yew (*Taxus baccata*) and used in modern medicine. The chemical *taxol*, found in the bark of some yews, was discovered to inhibit cell growth and division. It was therefore put to use in chemotherapy, halting the production of cancer cells. In addition, a homoeopathic tincture is made of young shoots. The berry flesh has been used by herbalists to treat a variety of ailments including cystitis, headache and neuralgia.

Recalling the text "From Me comes your fruit" (Hos 14:8), how can you become a source of healing for others?

6. Trunk

The physical strength of the yew's trunk has a vital role in holding the tree together through storm and calm. It symbolizes our need for a sturdy principle of unity or focus that unites the different parts of our life into a wholeness and completeness. It represents our need for an inner strength and an over-arching vision that creates a center to our life which can often be dissipated, as we are pulled this way and that by competing demands.

What brings unity and focus to the fragments of your life, holding diverse elements together? What is your personal uniting vision that encapsulates your vocation, the trunk of your life?

7. Bark

The yew's bark conveys a sense of beauty and mystery. The yew bark embodies the ages of the tree's history and character, all concealed behind a graceful posture. It exudes serenity and inspires deep and insightful thinking in all that encounter it.

Two types of questions suggest themselves:

Firstly, how would you describe the present texture of your soul? Does it feel vulnerable, flaky, malleable, crumbly? Are there sharp edges? Any brittle or fragile parts? Are there parts that feel cracked or scarred, weathered, hardened or flaky? Does your soul feel in any way smooth or gritty? How would you describe the bark of your soul?

Secondly, what persona do you present to the world? What do you reveal about yourself—and conceal? What do you feel about any possible woundedness, vulnerabilities, cracks or fractures? What kind of bark do you want to show to the world?

7 IMPERATIVES

1. Welcome Others

Yew hedges are incredibly dense, offering protection and nesting opportunities for many birds. In addition, I have noticed in Borderlands' garden a pattern of co-existence. One yew lives happily with a climbing rose, while another co-habits a space with hawthorn. Both live contentedly enough together as they get intertwined. There doesn't seem to be a sense of argument or competition!

How prepared are you to open your personal space or home to others? Do you see others as intrusions to be resented or blessings to be welcomed?

2. Safeguard Solitude

Yews stand alone, and in the Celtic tradition have been associated in their aloneness as with strong attributes such as power, honor, strength, and leadership. The Celtic tradition also recognizes the yew as having more mysterious traits, including silence, holiness, and introspection.

How can you combine in your life both hospitality and solitude, being alone and welcoming others, silence and talking, action and contemplation?

3. Stay Adaptable

Yew timber is not only incredibly strong and durable but also very adaptable. Traditionally, the wood was used in turnery making tool handles, cabinetry, carvings, musical instruments (lutes), and a great variety of turned objects. Among the hardest of all softwood species, it is possible to use yew for all manner of quality furniture and woodworking projects. It is also springy—literally flexible—long used for bows in archery. One of the world's oldest surviving wooden artefacts is fashioned from yew—the Clacton Spear estimated to be around an astonishing 420,000 years old!

How can you combine in your own life both sensitivity to others and to situations, and steadfastness? How can you be both strong yet responsive? How much "give" is there in you? In what circumstances are you inflexible or immovable?

4. Accept Pruning and Reshaping

Unlike most evergreens, a yew takes well to even harsh pruning because of its ability to regenerate new buds from old wood. Grown as a bush it is the favorite subject for the art of topiary and does not object to being regularly reshaped. Recall the words of John 15 (in relation to the vine) "my Father is the gardener. He cuts off every branch in me that bears no fruit, while every branch that does bear fruit he prunes so that it will be even more fruitful." (John 15:1b-2, NIV).

How do you respond to divine pruning or discipline (see Hebrews 12:5-11)? Are there any unruly parts of you or bits out of control that need the Gardener's knife? How has your life been re-shaped in recent years? But be sure also, as you reflect on this, to rejoice in your God-given beauty and dignity!

5. Embrace Paradox

It is growth itself that makes yew trees vulnerable as they can snap under their own weight or keel over in storms. But this rarely is the end. A yew can turn weakness or decay into health by allowing fungal infections to eat up its heartwood. This will leave a hollow tree but the tensile strength of the remaining wood continues to support the heavy crown of leaves. And as we noted, branches also loop down under their own weight until they touch the ground and set root. This evokes Paul's testimony: "He said to me, 'My grace is sufficient for you, for power is made perfect in weakness.' So, I will boast all the more gladly of my weaknesses, so that the power of Christ may dwell in me. Therefore I am content with weaknesses, insults, hardships, persecutions, and calamities for the sake of Christ; for whenever I am weak, then I am strong" (2 Cor 12: 9,10).

How do you live with the paradoxes in your life? Do you allow them to be life-giving?

6. Live the Easter Mystery

The branches of yew trees have traditionally been carried in Palm Sunday processions and used to decorate churches on Good Friday. They seem to have a fundamentally paschal character—dying to live. I've noticed that orangery dying leaves stay on the branches while fresh green shoots emerge each spring. Yews regenerate, we noted, not only by their drooping branches becoming rooted, but also inwardly, at their very core. Amazingly, whilst the center of a yew is dying, a branch may put down a root into the decaying leaf mold material inside the hollow trunk. The tree will regenerate! In this way a hollow yew is able to regenerate itself by producing new roots from its center. As these roots grow down into the ground they feed and strengthen the ageing tree, stabilizing it and prolonging its survival, enabling the tree to continue life long after many other trees would have perished. With this exceptional quality, it is understandable that the yew was revered as a symbol of long life, rebirth and regeneration. As old branches die new life can form within them. A tree that looks old and withered is constantly renewing itself, resurrecting itself. As a result, the trunks can grow to a massive diameter.

Recall Christ's words: "If any want to become my followers, let them deny themselves and take up their cross and follow me. For those who want to save their life will lose it, and those who lose their life for my sake, and for the sake of the gospel, will save it" (Mark 8:34–35).

What needs to die in you so that fresh expressions of divine new life might emerge?

7. Stay Evergreen!

Shimmering green yews abide as a symbol of hope and constancy through long grey winter days. They do not shed their leaves like the skeletal deciduous, but just keep on being themselves through thick and thin. So they have often been used for home decorations both in pagan midwinter solstice and Christmas celebrations.

How can you remain evergreen? The color evergreen bespeaks constancy and faithfulness, stickability and resilience in tough situations. How can we remain, in the striking words of the psalmist, "still full of sap still green"?

> In old age they still produce fruit;
> > they are always green and full of sap,
> > showing that the Lord is upright;
> > > he is my rock, and there is no unrighteousness in him. (Ps 92:15)

The secret of staying open to rejuvenation and renewal is the synergy of the disciple with the Gardener:

> Work out your own salvation with fear and trembling; for it is God who is at work in you, enabling you both to will and to work for his good pleasure. (Phil 2:12,13)

> For we are laborers together with God: ye are God's husbandry (1 Cor 3:9, AV)

What strategies for growth and ongoing renewal are emerging for you, as you reflect in this retreat on the replanting of your soul?

Staying Ever Open to the Call of God

John Henry Newman (1801–90), recognized as a saint in 2019, encourages us—despite our strategies and plans—to stay ever open to the call of God, however it may be mediated to us:

> For in truth we are not called once only, but many times; all through our life Christ is calling us. He called us first in Baptism; but afterwards also; whether we obey His voice or not, He graciously calls us still . . . He calls us on from grace to grace, and from holiness to holiness, while life is given us. Abraham was called from his home, Nathaniel from his retreat; we are all in course of calling, on and on, from one thing to another, having no resting-place, but mounting towards our eternal rest, and obeying one command only to have another put upon us . . .
>
> It were well if we understood this; but we are slow to master the great truth, that Christ is, as it were, walking among us, and by His hand, or eye, or voice, bidding us follow Him. We do not understand that His call is a thing which takes place now. We think it

took place in the Apostles' days; but we do not believe in it, we do not look out for it in our own case ...

What happens to us in providence is in all essential respects what His voice was to those whom He addressed when on earth: whether he commands us by a visible presence, or by a voice, or by our consciences, it matters not, so that we feel it to be a command. If it is a command, it may be obeyed or disobeyed ...

And these divine calls are commonly, from the nature of the case, sudden now, and as indefinite and obscure in their consequences as in former times ... One is going on as usual; he comes home one day, and finds a letter, or a message, or a person, whereby a sudden trial comes upon him, which, if met religiously, will be the means of advancing him to a higher state of religious excellence ...

Again, perhaps something occurs to force us to take a part for God or against Him ... Some tempting offer is made us; or some reproach or discredit threatened us; or we have to determine and avow what is truth and what is error ... That little deed, suddenly exacted of us, almost suddenly resolved on and executed, may be as though a gate into the second or third heaven—an entrance into a higher state of holiness, and into a truer view of things than we have thitherto taken.

Or again, we get acquainted with someone whom God employs to bring before us a number of truths which were closed on us before; and we but half understand them, and but half approve of them; and yet God seems to speak in them, and Scripture to confirm them. This is a case which not infrequently occurs, and it involves a call "to follow on to know the Lord."

Or again, we may be in the practice of reading Scripture carefully, and trying to serve God, and its sense may, as if suddenly, break upon us , in a way it never did before. Some thought may suggest itself to us which is the key to a great deal in Scripture, or which suggests a great many other thoughts. A new light may be thrown on the precepts of our Lord and His Apostles. We may be able to enter into the manner of life of the early Christians, as recorded in Scripture, which before was hidden from us, and into the simple maxims on which Scripture bases it. We may be led to understand that it is very different from the life which men lead now. Now knowledge is a call to action: an insight into the way of perfection is a call to perfection.[3]

---

3. Przywara, *Heart of Newman*, 198–202.

I have Come that You may have Life and Life in All its Fulness

"I came that they may have life, and have it abundantly" (10:10). This is life in all its exuberance. The Greek word *perissos* means above measure, more than average, above the common, extraordinary, more than sufficient, with a surplus. The *Amplified Bible* puts it: "I came that they may have and enjoy life, and have it in abundance, to the full, till it overflows." Jesus offers life in its fullest measure, life that is full and good. The *Message* gives us: "I came so they can have real and eternal life, more and better life than they ever dreamed of." The *New Life Version* renders this: "I came so they might have life, a great full life." Above all, the trees we have sought to plant afresh in our soul during this retreat are symbols of the divine life growing and fruiting within us:

- The fig tree beckoned us to open up spaces in our lives where we can create stillness and receptivity to the Divine.
- The oaks invited us to celebrate God's providence and guidance in our life as we celebrate the journey thus far.
- The willow allowed us to "hang up our harps" and let go of attachments, burdens and sorrows—and also identify possibilities for rejoicing.
- The vine paradoxically called us to stillness and to abide and settle in God's grace, while grapes crushed spoke to us of sacrificial ministry.
- The apple tree not only summoned us back to Eden and forwards to heaven but invited us to delight fully in Christ's grace here and now.
- The cedar summoned us to reflect on our dignity and also on our availability to God.
- The olive reminded us of the need to embrace the life-giving paradox of the Cross.
- The yew tree held out to us the possibility of unending renewal.

PRAYER EXERCISE

As you reflect on the two sets of seven challenges presented by the yew tree, slowly use these words spoken by wisdom personified—and apply them to yourself. Let them express the prayer that your personal vocation will continue to flourish, open out and bear amazing, perhaps unexpected, fruit!

> I took root in an honored people,
> > in the portion of the Lord, his heritage.
> I grew tall like a cedar in Lebanon,
> > and like a cypress on the heights of Hermon.
> I grew tall like a palm tree in En-gedi,
> > and like rosebushes in Jericho;
> like a fair olive tree in the field,
> > and like a plane tree beside water I grew tall.
> Like cassia and camel's thorn I gave forth perfume,
> > and like choice myrrh I spread my fragrance ...
> Like a terebinth I spread out my branches,
> > and my branches are glorious and graceful.
> Like the vine I bud forth delights,
> > and my blossoms become glorious and abundant fruit.
> Come to me, you who desire me,
> > and eat your fill of my fruits.
> For the memory of me is sweeter than honey,
> > and the possession of me sweeter than the honeycomb.
> Those who eat of me will hunger for more,
> > and those who drink of me will thirst for more. (Sir 24:13–21)

Allow God to address you and send you forth through these words:

> You shall go out in joy,
> > and be led forth in peace;
> the mountains and the hills before you
> > shall break forth into singing,
> > and all the trees of the field shall clap their hands.
> Instead of the thorn shall come up the cypress;
> > instead of the brier shall come up the myrtle;
> and it shall be to the LORD for a memorial,
> > for an everlasting sign which shall not be cut off. (Isa 55:12–12)

Bibliography

Armstrong, Regis J., et al., eds. *Francis of Assisi Early Documents: Vol. 2, The Founder.* New York: New City, 2000.
Avis, Paul. *God and the Creative Imagination: Metaphor, Symbol and Myth in Religion and Theology.* London: Routledge, 1999.
Barry, William A. and Connolly, William J. *The Practice of Spiritual Direction.* New York: Seabury, 1982.
Blaiklock, E. M. and Keys, A. C., trans. *The Little Flowers of St Francis.* London: Hodder and Stoughton, 1985.
Bonaventure. *The Soul's Journey into God; The Tree of Life; The Life of St Francis.* Translated by Ewert Cousins. New York: Paulist, 1978.
Bowie, Fiona and Davies, Oliver, eds. *Hildegard of Bingen: An Anthology.* London: SPCK, 1990.
Brock, Sebastian, trans. *St. Ephrem the Syrian: Hymns on Paradise.* Crestwood, NY: St. Vladimir's Seminary, 1990.
———. *The Luminous Eye: The Spiritual World Vision of Saint Ephrem the Syrian.* Kalamazoo, Michigan: Cistercian, 1992.
———. *Isaac of Nineveh: The Second Part.* Leuven: Peeters, 1995.
Bryant, Christopher. *Journey to the Centre.* London: Darton, Longman and Todd, 1987.
Campbell, Joseph. *Thou Art That: Transforming Religious Metaphor.* Novato, CA: New World Library, 2013.
Colledge, Edmund and Walsh, James, trans. *Julian of Norwich: Showings.* New York: Paulist, 1978.
Farber, Zev. "The Mystical Ritual of Hoshana Rabbah: Summoning God" https://www.thetorah.com/article/the-ritual-of-hoshana-rabbah.
Foster, Richard. *Prayer: Finding the Heart's True Home.* London: Hodder and Stoughton 1992.
Fox, Matthew. *Illuminations of Hildegard of Bingen.* Rochester, Vermont: Bear and Company, 1985.
———. *Hildegard of Bingen's Book of Divine Works.* Sante Fe, New Mexico: Bear and Company, 1987.
Fuhrkotter. Adelgundis, ed. *Hildegardis Scivias.* Turnhaut: Corpus Christianorum Continuation Mediaevalis, 1978.
Goddard, Ben. "Churchyard yew tree dates back 3,000 years." https://www.herefordtimes.com/news/14095704.churchyard-yew-tree-dates-back-3000-years/.

Hauerwas, Stanley and Willimon, William H. *Resident Aliens: A Provocative Christian Assessment of Culture and Ministry for People who know that Something is Wrong*. Nashville: Abingdon, 1989.

Hillel, Daniel. *The Natural History of the Bible: An Environmental Exploration of the Hebrew Scriptures*. New York: Columbia University Press, 2006.

Hopkins, Gerard Manley *Poems and Prose*. London: Penguin, 1985.

Humphreys, Caroline. *From Ash to Fire: A Contemporary Journey through the Interior Castle of Teresa of Avila*. New York: New City, 1992.

Inge, Denise. *Happiness and Holiness: Thomas Traherne and His Writings*. Norwich: Canterbury, 2008.

Jordan, Patricia. *An Affair of the Heart: a Biblical and Franciscan Journey*. Leominster: Gracewing, 2008.

Kavanaugh, Kieran and Rodriguez, Otilio, trans. *Teresa of Avila: The Interior Castle*. New York: Paulist, 1979.

Kennedy, Charles William. *The Poems of Cynewulf*. New York: E. P. Dutton and Co., 1910.

Kinver, Kinver. "World losing Battle Against Deforestation." https://www.bbc.co.uk/news/science-environment-49679883.

Lakoff, George and Johnson, Mark. *Metaphors We Live By*. Chicago: University of Chicago Press,1980.

Mayes, Andrew D. *Learning the Language of the Soul: A Spiritual Lexicon*. Collegeville MN; Liturgical, 2016.

———. *Sensing the Divine: John's Word Made Flesh*. Abingdon: BRF, 2019.

———. *Diving for Pearls: Discovering the Depths of Prayer with Isaac the Syrian*. Collegeville MN: Cistercian, 2021.

Merton, Thomas. *A Search for Solitude: Pursuing the Monk's True Life*. The Journals of Thomas Merton, Volume 3: 1952-1960. New York: HarperOne, 2009.

———. *Contemplation in a World of Action*. London: George Allen and Unwin, 1971.

———. *Sign of Jonas*, New York: HarperOne, 2002.

Moltmann, Jurgen. *The Spirit of Life: An Universal Affirmation*. Minneapolis: Fortress, 1992.

Muggeridge, Kitty, trans. *The Sacrament of the Present Moment: Jean-Pierre de Caussade*. London: Fount, 1996.

Murray, Robert. *Symbols of Church and Kingdom: A Study in Early Syriac Tradition*. London: T and T Clark International, 2006.

Obbard, Elizabeth R. *Through Julian's Windows: Growing into Wholeness with Julian of Norwich*. Norwich: Canterbury, 2008.

Peers, Allison, trans. *St Teresa of Avila: Interior Castle*. London: Sheed and Ward, 1974.

Phillips, C., ed. *Gerald Manley Hopkins: The Major Works*. Oxford: University Press, 1986.

Przywara, Erich, ed. *The Heart of Newman: Parochial and Plain Sermons*. Wheathampstead: Anthony Clarke, 1963.

Rolheiser, Ronald. *Seeking Spirituality*. London: Hodder and Stoughton, 1998.

Rotzetter, Anton, et al. *Gospel Living: Francis of Assisi Yesterday and Today*. New York: Franciscan Institute, 1994.

Society of Martha and Mary. *Affirmation and Accountability*. Dunsford: Society of Martha and Mary, 2002.

Soskice, Janet M. *Metaphor and Religious Language*. Oxford: Clarendon, 1985.

Spurgeon, Charles. "The Trees in God's Court", *Metropolitan Tabernacle Pulpit Volume 23*. https://www.spurgeon.org/collection/metropolitan-tabernacle-pulpit-volume-23/.

Stoutzenberger Joseph M. and Bohrer, John D. *Praying with Francis of Assisi*. Winona, Minnesota: Saint Mary's, 1989.
Thomas, James H. *The Pilgrim's Progress in Today's English*. Eastbourne: Victory, 1972.
Traherne, Thomas. *Centuries of Meditations*. London: Bertram Dobell, 1908.
Uhlein, Gabriele. *Meditations with Hildegard of Bingen*. Sante Fe, New Mexico: Bear and Company, 1983.
Wade, Gladys L., ed. *The Poetical Works of Thomas Traherne*. London: Dobell, 1932.
Wensinck Arent Jan, trans. *Mystical Treatises by Isaac of Nineveh*. Amsterdam: Nieuwe Reeks, 1923.
Wesley, John. *Sermons on Several Occasions*. Grand Rapids, MI: Christian Classics Ethereal Library.
Wiederkehr, Macrina. *Seasons of Your Heart: Prayers and Reflections*. San Francisco: HarperSanFrancisco, 2004.
Williams, Rowan. *The Edge of Words: God and the Habits of Language*. London: Bloomsbury, 2014.
Wills, Simon. *A History of Trees*. Barnsley: Pen and Sword, 2018. Windeatt B. A., trans. *The Book of Margery Kempe*. Harmondsworth: Penguin, 2000.
Woodland Trust. woodlandtrust.org.uk/plant-trees/schools-and-communities/queens-green-canopy/.
Wren, Brian. *What Language Shall I Borrow? God –Talk in Worship*. London: SCM, 1989.

Ingram Content Group UK Ltd.
Milton Keynes UK
UKHW021554110723
424903UK00006B/32